ALLE·ZEIT·WACH
1842

Wang Yuan

Diophantine Equations and Inequalities in Algebraic Number Fields

Springer-Verlag
Berlin Heidelberg New York London Paris
Tokyo Hong Kong Barcelona

Prof. Wang Yuan
Institute of Mathematics
Academia Sinica
Beijing
CHINA

Mathematics Subject Classification (1980): 10B30, 10B45, 10C10, 10D30, 10F45, 10G05, 10G10, 10J06, 10J10, 10L02, 12A40

ISBN 3-540-52019-8 Springer-Verlag Berlin Heidelberg New York
ISBN 0-387-52019-8 Springer-Verlag New York Berlin Heidelberg

Library of Congress Cataloging-in-Publication Data
Wang, Yuan, 1930- Diophantine equations and inequalities in algebraic number fields / Wang Yuan. p. cm. Includes bibliographical references and index.
ISBN 0-387-52019-8 (U.S.)
1. Fields, Algebraic 2. Diophantine analysis. 3. Inequalities (Mathematics) I. Title.
QA247.W345 1991 512'.72 - dc20 90-23989

Printed in the United States of America

41/3140-543210 - Printed on acid-free paper

Contents

Introduction

The circle method has its genesis in a paper of Hardy and Ramanujan (see [Hardy 1]) in 1918 concerned with the partition function and the problem of representing numbers as sums of squares. Later, in a series of papers beginning in 1920 entitled "some problems of 'partitio numerorum'", Hardy and Littlewood (see [Hardy 1]) created and developed systematically a new analytic method, *the circle method in additive number theory.* The most famous problems in additive number theory, namely Waring's problem and Goldbach's problem, are treated in their papers. The circle method is also called the Hardy-Littlewood method.

Waring's problem may be described as follows: For every integer $k \geq 2$, there is a number $s = s(k)$ such that every positive integer N is representable as

$$N = x_1^k + \cdots + x_s^k, \tag{1}$$

where x_i are non-negative integers. This assertion was first proved by Hilbert [1] in 1909.

Using their powerful circle method, Hardy and Littlewood obtained a deeper result on Waring's problem. They established an asymptotic formula for $r_s(N)$, the number of representations of N in the form (1), namely

$$r_s(N) = \Gamma\left(1 + \frac{1}{k}\right)^s \Gamma\left(\frac{s}{k}\right)^{-1} \mathfrak{S}(N) N^{s/k-1}(1 + o(1)), \tag{2}$$

provided that $s \geq (k-2)2^{k-1} + 5$. Here $\mathfrak{S}(N)$ is the so-called singular series which has a positive lower bound independent of N. Let $G(k)$ be the least value of s such that every large N is representable in the form (1). Then (2) implies that

$$G(k) \leq (k-2)2^{k-1} + 5. \tag{3}$$

The proof of (2) can be sketched as follows: Let

$$g(z) = \sum_{N=1}^{\infty} r_s(N) z^N = \left(\sum_{x=1}^{\infty} z^{x^n}\right)^s, \qquad |z| < 1.$$

By Cauchy's formula, we have

$$r_s(N) = \frac{1}{2\pi i} \int_\Gamma g(z) z^{-N-1} dz,$$

where Γ is the circle $|z| = \rho$ with $0 < \rho < 1$. Now this circle Γ is divided into two parts. Roughly speaking, one part contains the disjoint arcs with centres $\rho e(a/q)$, where $(a,q) = 1$ and q is "small". The union $\mathfrak{M}$ of these arcs is called the "major arcs". The remaining part $\mathfrak{m}$ of Γ with respect to $\mathfrak{M}$ is called the "minor arcs". We call this division of Γ into $\mathfrak{M}$ and $\mathfrak{m}$ a *Farey division.* The integral over $\mathfrak{M}$ then gives the main term of $r_s(N)$. The fact that the integral over $\mathfrak{m}$ is of a lower order follows from Weyl's inequality. This inequality was given in a less explicit form in Weyl's famous memoir of 1916 on the uniform distribution of sequences of numbers modulo 1; see Weyl [1].

In 1928, Vinogradov (see [Vinogradov 3]) introduced a number of notable refinements on the circle method, one of which is to replace $g(z)$ by the finite exponential sum

$$h(z) = \left(\sum_{x=1}^{T} e(x^k z)\right)^s, \qquad T = [N^{1/k}].$$

For integral values a, the orthogonality relation

$$\int_0^1 e(az)\,dz = \begin{cases} 1, & \text{if } \ a = 0, \\ 0, & \text{if } \ a \neq 0, \end{cases}$$

gives

$$r_s(N) = \int_0^1 h(z)\,e(-zN)\,dz.$$

Further remarkable progress on the circle method and related problems was made by Vinogradov (see [Vinogradov 3]) in the 1930's and these are based on his ingeneous method for the estimation of exponential sums. In particular he proved that the asymptotic formula (2) holds for

$$s \geq 2k^2(2\log k + \log\log k + 2{\cdot}5), \qquad k \geq 11 \tag{4}$$

(see [Vinogradov 3] and [Hua 3]), and that

$$G(k) \leq 2k\log k + 2k\log\log k + 12k; \tag{5}$$

see [Vinogradov 3] and Karatsuba [1]. Very recently T. D. Wooley [1] has made the following substantial improvement

$$G(k) \leq k\log k + k\log\log k + O(k).$$

For small k, Hua's inequality

$$\int_0^1 \left|\sum_{x=1}^{T} e(x^k z)\right|^{2^k} dz \ll T^{2^k - k + \epsilon} \tag{6}$$

shows that (2) holds for

$$s \geq 2^k + 1; \tag{7}$$

see [Hua 4]. This has been improved recently by Vaughan [1,2] to $s \geq 2^k$ $(k \geq 3)$ and by Heath-Brown [3] to $s \geq \frac{7}{8}2^k + 1\,(k \geq 6)$.

A new elementary proof of Hilbert's theorem was given by Linnik [1], and it is based on Schnirelman's method on the density of sequences of natural numbers; see Khintchine [1] and [Hua 1]. It has been pointed out by Davenport [Davenport 1] that the ideas of his proof were undoubtedly suggested by certain features of the Hardy-Littlewood method, and in particular, by Hua's inequality.

Besides the estimation for the singular series, there is no essential difficulty in extending the circle method to deal with the more general equation of additive type, say

$$N = f_1(x_1) + \cdots + f_s(x_s), \tag{8}$$

where $f_i(x)$ are integral valued polynomials of degree k with positive leading coefficients; see [Hua 4]. We are interested in particular in the so-called additive equation

$$a_1 x_1^k + \cdots + a_s x_s^k = 0, \tag{9}$$

where a_i are natural numbers.

Using their variant of the circle method, Davenport and Heilbronn [1] were able to treat the diophantine inequality for additive forms. They proved that an inequality

$$\left|\alpha_1 x_1^2 + \cdots + \alpha_5 x_5^2\right| < 1$$

with real $\alpha_1, \ldots, \alpha_5$ not all of the same sign, has a non-trivial solution. The same conclusion was later shown to be true for a general indefinite quadratic form in at least 21 variables; see Davenport and Ridout [1]. One of the outstanding open problems is to show that 3 variables will suffice if the coefficients are not all in rational ratios. This problem was solved recently by Margulis [1,2].

It is difficult to deal with equations of non-additive type by the circle method. The first substantial progress on this was made by Brauer [1]. He showed that if F is a field and if, for every $k \geq 1$, there is a $c_1(k, F)$ such that equation (9) with $s \geq c_1$ and with coefficients in F has a non-trivial solution, i.e., a solution with at least one x_i not zero, then every form $f(\mathbf{x}) = f(x_1, \ldots, x_s)$ of degree k with coefficients in F in at least $s \geq c_2(k, F)$ variables possesses a non-trivial zero in F^s. Brauer's method may be described as a purely algebraic "diagonalisation" process. It does not apply to the rational field $\mathbb{Q}$ since additive equations of even degree may have no non-trivial solution in $\mathbb{Q}^s$. However Birch [1] (see also [Vaughan 1], Chapter 9) was able to show that if the hypothesis of Brauer's theorem is satisfied for all odd values of k, then the conclusion also holds for all forms of odd degree. So he has established that if $f(\mathbf{x})$ is a form of odd degree k with rational coefficients in $s \geq c_3(k)$ variables, then $f(\mathbf{x})$ has a non-trivial zero in $\mathbb{Z}^s$. Davenport [1] proved that $c_3(3) \leq 16$, and later Heath-Brown [1] reduced 16 to 10 for non-singular cubic forms.

Forms with real coefficients are more difficult to deal with. Schmidt [3] proved a remarkable theorem which states that if $f(\mathbf{x})$ is a form of odd degree k with real coefficients in $s \geq c_4(k)$ variables, then there is an integer point

$\mathbf{x} \in \mathbb{Z}^s$ with $\mathbf{x} \neq \mathbf{0}$ and $|f(\mathbf{x})| < 1$. It immediately yields Birch's theorem if the coefficients of $f(\mathbf{x})$ are integral. More precisely, Schmidt proved the following deeper result:

Let $f_i(\mathbf{x}) = f_i(x_1, \ldots, x_s)\,(1 \leq i \leq h)$ be forms of odd degrees $\leq k$ with real coefficients. Given a positive number E, however large, there is a constant $c_5(k, h, E)$ with the following property. Let $T \geq 1$ and $s \geq c_5$. Then there is a non-zero integer point $\mathbf{x} \in \mathbb{Z}^s$ with

$$|\mathbf{x}| \leq T \quad \text{and} \quad |f_i(\mathbf{x})| \ll T^{-E}|f_i|, \quad 1 \leq i \leq h, \tag{10}$$

where $|\mathbf{x}| = \max|x_i|$ and $|f|$ denotes the maximum value of the coefficients of f.

Suppose that $f_i(\mathbf{x})\,(1 \leq i \leq h)$ are forms with coefficients in $\mathbb{Z}$. From the above theorem the following result may be derived: For any $\epsilon > 0$, there is a constant $c_6(k, h, \epsilon)$ such that if $s \geq c_6$, then there exists a non-zero integer point $\mathbf{x} \in \mathbb{Z}^s$ with

$$|\mathbf{x}| \ll F^{\epsilon} \quad \text{and} \quad f_i(\mathbf{x}) = 0, \quad 1 \leq i \leq h, \tag{11}$$

where $F = \max(1, |f_i|)$.

The first step in the proof of Schmidt's theorem is to prove (11) for a single additive equation. Although the circle method is used in his proof, the treatment of the minor arcs is distinct from that for Waring's problem. The second step is to apply Davenport and Heilbronn's variant circle method to prove (10) for a single additive form by his new method of treating the "minor arcs" and the rational approximations to coefficients. Finally he established his theorem by his refined Brauer-Birch diagonalisation method. This is the background of the theory of diophantine equations and inequalities in algebraic number fields.

Siegel ([1–4] or [Siegel 1]) was interested in generalising the circle method to an arbitrary algebraic number field $\mathbb{K}$ of degree n, and he tried to solve the analogue of Waring's problem for $\mathbb{K}$ in 1922. Siegel noted that Waring's problem in algebraic number fields has a different character from that in the rational field as shown by the following example. Let $\mathbb{Q}(\sqrt{d})$ be a quadratic field with $d \equiv 2, 3 \pmod 4$. The integers of such a field are of the form $a + b\sqrt{d}$ with $a, b \in \mathbb{Z}$. The square of such an integer has an even second coefficient, and therefore an integer with an odd second coefficient is not a sum of squares. This led Siegel to consider the ring J_k generated by the k-th powers of the integers of $\mathbb{K}$, instead of the ring J of integers of $\mathbb{K}$. Siegel [1,2] first succeeded in dealing with the problem of decomposition into square numbers for $\mathbb{K}$ in 1922. Finally, in 1944–45, he (Siegel [3,4]) established his generalised circle method for $\mathbb{K}$ and proved a result on Waring's problem for $\mathbb{K}$ corresponding to (2) when $s \geq (2^{k-1} + n)kn$. Earlier, in the 1920's, he had not been able to overcome a difficulty in generalising the major arcs and minor arcs of the Farey division. Siegel's result was improved and generalised by Ayoub, Birch, Eda, Körner,

Mitsui, Peck, Stemmler, Subbarao, Tatuzawa and Wang in various aspects to the present (see Reference II).

Rieger [1,2] generalised Schnirelman's method and proved that, for any algebraic number field $\mathbb{K}$, the set of sums of $c_7(k)(= 8^{k-1}/2)$ k-th powers of totally non-negative integers of $\mathbb{K}$ always has a positive density in J_k. It is interesting that c_7 is independent of the degree of $\mathbb{K}$.

Let $\mathbb{K}$ be a purely imaginary algebraic number field of degree $2r$ with ring of integers J. Using Siegel's method, Peck [1] proved that, for every $k \geq 1$, there is a $c_8(k,r)$ such that the equation (9), with $s \geq c_8$ and with coefficients in J, always has a non-trivial solution in J^s. So it follows immediately by Brauer's theorem that any form $f(\mathbf{x}) = f(x_1, \ldots, x_s)$ of degree k with coefficients in J has a non-trivial zero in J^s if $s \geq c_9(k, \mathbb{K})$. By the combination of the methods of Schmidt and Siegel, we can prove the following result:

Let $f_i(\boldsymbol{\lambda}) = f_i(\lambda_1, \ldots, \lambda_s)\,(1 \leq i \leq h)$ be forms of degree $\leq k$ with complex coefficients. Given a positive number E, however large, there is a constant $c_{10}(k,h,r,E)$ with the following property. Let $T \geq 1$ and $s \geq c_{10}$. Then there exists a non-zero point $\boldsymbol{\lambda} \in J^s$ satisfying

$$\|\boldsymbol{\lambda}\| \leq T \quad \text{and} \quad \left|f_i(\boldsymbol{\lambda})\right| \ll T^{-E}|f_i|, \quad 1 \leq i \leq h, \tag{13}$$

where $\|\boldsymbol{\lambda}\|$ denotes the maximum absolute value of its components and their conjugates.

The achievements on the Hardy-Littlewood method and its related problems are given in many books; see Reference I. In particular Vaughan's tract [Vaughan 1] gives an excellent account of the major achievements of the circle method to date and contains a comprehensive bibliography. Schmidt's method on diophantine equations and inequalities in the rational case is also given in details in the recent book of Baker; see [Baker 1], Chapters 11, 13 and 14. So only a small number of the related work on the Hardy-Littlewood method and its applications is listed in this book; see Reference II.

The purpose of this book is to give an account of Siegel's generalised circle method and its applications to Waring's problem and additive equations. We also discuss Schmidt's method on diophantine equations and inequalities in many variables in algebraic number fields. Chapter 1 deals with the Hardy-Littlewood method and its application to Waring's problem, where (2) will be established for $s \geq 2^k + 1$. The estimation of exponential sums plays an important role in number theory; this is particularly so in the circle methods of Hardy-Littlewood and Siegel, and it will be dealt with in Chapters 2–4. Chapters 5–8 contain an exposition of Siegel's method and its applications to Waring's problem and additive equations including the Schnirelman-Linnik-Hua-Rieger method on Waring's problem. Finally, Chapters 9–11 deal with Schmidt's method on small solutions for diophantine equations in many variables and his method on diophantine inequalities for forms in algebraic number fields.

Since there are many important results on the Hardy-Littlewood method which so far have no analogues in algebraic number fields, there is hardly a need for us to suggest open problem for investigation. But it might have been useful to devote some space to the work of equations with small degrees, the small solutions of congruences and the Vinogradov-Goldbach theorem in number fields. However, this would considerably lengthened this book. The interested reader is referred to Reference II.

It is only assumed that the reader has a familiarity with elementary number theory and the classical theory of algebraic number fields. Such background for the reader can be found in excellent books by Hecke [Hecke 1] and Hua [Hua 1].

In conclusion, I wish to express my indebtedness to Professor W. M. Schmidt for his helpful suggestions and advice on my work. I also offer thanks to the referees, and Dr. P. Shiu in particular, for their suggestions on improvements, and to Springer-Verlag for their help during the course of publication.

June 1990

Wang Yuan

Notation

For a set F, we let F^s denote the set of s-tuples $\mathbf{x} = (x_1, \ldots, x_s)$ with $x_i \in F$, $1 \le i \le s$. The rational field, the real field and the ring of integers are denoted by $\mathbb{Q}$, $\mathbb{R}$ and $\mathbb{Z}$ respectively. The n-dimensional Euclidean space is denoted by E_n, and we write U_n for the unit cube $\{\mathbf{x} : 0 \le x_i \le 1;\ 1 \le i \le n\}$.

We let $\mathbb{K}$ denote an algebraic number field of degree n over $\mathbb{Q}$. We then let $\mathbb{K}^{(\ell)}$ $(1 \le \ell \le r_1)$ be the real conjugates while $\mathbb{K}^{(m)}$ and $\mathbb{K}^{(m+r_2)}$ $(r_1 + 1 \le m \le r_1 + r_2)$ are the complex conjugates of $\mathbb{K}$, where $r_1 + 2r_2 = n$. Throughout this book the indices ℓ and m are over the sets on integers cited here. The letter J denotes the ring of all integers of $\mathbb{K}$.

For $\gamma \in \mathbb{K}$, $\gamma^{(i)}$ $(1 \le i \le n)$ denote the conjugates of γ, where $\gamma^{(i)} \in \mathbb{K}^{(i)}$, $1 \le i \le n$. For $\gamma_j \in \mathbb{K}$ and $x_j \in \mathbb{R}$, $1 \le j \le n$, we set $\lambda = \sum_{1 \le j \le n} x_j \lambda_j$ and define $\lambda^{(i)} = \sum_{1 \le j \le n} x_j \gamma_j^{(i)}$, $1 \le i \le n$. We also let

$$\exp(x) = e^x, \quad e(x) = \exp(2\pi i x), \quad S(\lambda) = \sum_{i=1}^{n} \lambda^{(i)}, \quad E(\lambda) = e\big(S(\lambda)\big),$$

$$N(\lambda) = \prod_{i=1}^{n} \lambda^{(i)}, \qquad \|\lambda\| = \max \big|\lambda(i)\big|, \qquad \|\boldsymbol{\lambda}\| = \max_j \|\lambda_j\|$$

with a vector $\boldsymbol{\lambda} = (\lambda_1, \ldots, \lambda_s)$.

We use $w_1, \ldots, w_n$ to denote an integral basis of $\mathbb{K}$, i.e., a basis of J; also δ is the different (ground ideal), and D is the absolute value of the discriminant of $\mathbb{K}$. The numbers $\rho_1, \ldots, \rho_n$ form a basis of δ^{-1} such that

$$S(\rho_i w_j) = \begin{cases} 1, & \text{if } \ i = j, \\ 0, & \text{if } \ i \ne j. \end{cases}$$

We shall write, for real x_i and y_i,

$$\begin{aligned} \xi &= x_1\rho_1 + \cdots + x_n\rho_n, & \eta &= y_1 w_1 + \cdots + y_n w_n, \\ dx &= dx_1 \cdots dx_n, & dy &= dy_1 \cdots dy_n. \end{aligned}$$

A number γ of $\mathbb{K}$ is said to be *totally non-negative* if $\gamma^{(\ell)} \ge 0$. The set of all totally non-negative integers of $\mathbb{K}$ is denoted by P.

We use $P(T)$ to denote the set of $\mathbf{y} \in \mathbb{R}^n$ satisfying

$$0 \le \eta^{(\ell)} \le T, \qquad \big|\eta^{(m)}\big| \le T,$$

and $M(T)$ is the set of $\mathbf{y}$ with $\|\eta\| \le T$. Sums with λ running over all integers of $P(T)$ and $M(T)$ are denoted by $\sum_{\lambda \in P(T)}$ and $\sum_{\lambda \in M(T)}$ respectively.

The Gothic letters $\mathfrak{a}$, $\mathfrak{b}$, ... denote integral or fractional ideals, and $\wp, \wp_1, \ldots$ the prime ideals of $\mathbb{K}$. We write $N(\mathfrak{a})$ for the norm of $\mathfrak{a} = (\alpha_1, \ldots, \alpha_s)$, the ideal generated by $\alpha_1, \ldots, \alpha_s$.

For ideals $\mathfrak{a}$ and $\mathfrak{b}$, $\mathfrak{a}|\mathfrak{b}$ means that $\mathfrak{a}$ divides $\mathfrak{b}$; otherwise we write $\mathfrak{a}\nmid\mathfrak{b}$, and $\mathfrak{a}^i\|\mathfrak{b}$ means $\mathfrak{a}^i|\mathfrak{b}$, but $\mathfrak{a}^{i+1}\nmid\mathfrak{b}$. A sum with λ running over a complete residue system modulo $\mathfrak{a}$ is denoted by $\sum_{\lambda(\mathfrak{a})}$.

The letter k is an integer greater than 1, and c is used to denote an absolute constant, while $c(f,\ldots,g)$ is a positive constant that depends on $f,\ldots,g$, but not always having the same value in different occurances. The letters $\epsilon,\epsilon_1,\ldots$ are preassigned positive numbers less than 1. The symbols $\ll$ (or O) and o have their usual meanings, and the constants implicit in $\ll$, $\gg$ (or O) and o often depend on k, $\mathbb{K}$ and ϵ.

Chapter 1

The Circle Method and Waring's Problem

1.1 Introduction

In this chapter we introduce the circle method of Hardy and Littlewood and its application to Waring's problem in the rational field $\mathbb{Q}$.

For $N > 1$, let $r_s(N)$ be the number of solutions to the equation

$$N = x_1^k + \cdots + x_s^k, \tag{1.1}$$

where $x_i\,(1 \le i \le s)$ are non-negative integers. Let $G(k)$ be the least integer s such that (1.1) is soluble for every sufficiently large N.

Theorem 1.1. *Suppose that* $s \ge 2^k + 1$. *Then*

$$r_s(N) = \mathfrak{S}(N)\Gamma\left(1+\frac{1}{k}\right)^s \Gamma\left(\frac{s}{k}\right)^{-1} N^{s/k-1}(1+o(1))$$

where

$$\mathfrak{S}(N) = \sum_{q=1}^{\infty}\sum_{a=1}^{q}{}' \left(\frac{1}{q}\sum_{b=1}^{q} e\left(\frac{ab^k}{q}\right)\right)^s e\left(-\frac{aN}{q}\right) \ge c_1(k,s) > 0.$$

Here Σ' denotes a sum with $(a,q) = 1$, and the expression $\mathfrak{S}(N)$ is called the *singular series*. From Theorem 1.1 we derive immediately

Theorem 1.2. $G(k) \le 2^k + 1$.

1.2 Farcy Division

Let

$$\delta = \frac{1}{2k}, \quad T = \left[N^{1/k}\right], \quad t = \left[T^{1-\delta}\right] \quad \text{and} \quad h = \left[N^{k-1+\delta}\right].$$

According to Dirichlet's approximation theorem (see, for example, [Hua 1; §6.10]), for each $\alpha \in [0,1)$, there exists a fraction a/q such that

$$\left|\alpha - \frac{a}{q}\right| \le \frac{1}{qh}, \qquad (a,q) = 1. \tag{1.2}$$

For $1 \le a \le q \le t$ with $(a,q) = 1$, we let $\mathfrak{M}(a,q)$ be the set of points

$$\left\{\alpha : \left|\alpha - \frac{a}{q}\right| \le \frac{1}{qh}\right\};$$

we sometimes write this in the form $\max(h\,|\alpha - a/q|, 1/t) \le 1/q$, which is in agreement with the notation to be introduced in Chapter 5. We also use $\mathfrak{M}(1,1)$ to denote the union of the two intervals $[0, 1/h]$ and $[1 - 1/h, 1)$. The sets $\mathfrak{M}(a,q)$ are pairwise disjoint. In fact, if $a/q \ne a'/q'$ and $\mathfrak{M}(a,q) \cap \mathfrak{M}(a',q') \ne \phi$, then

$$\frac{1}{q'h} > \frac{1}{h}\left(\frac{1}{q} + \frac{1}{q'}\right) \ge \left|\frac{a}{q} - \frac{a'}{q'}\right| \ge \frac{1}{qq'}$$

so that $h < q \le t$, which is impossible. The union

$$\mathfrak{M} = \bigcup \mathfrak{M}(a,q)$$

forms the major arcs, and the complementary set $\mathfrak{m}$ of $\mathfrak{M}$ with respect to $[0,1)$ is called the minor arcs. The division of $[0,1)$ into $\mathfrak{M}$ and $\mathfrak{m}$ depends on (h,t), and we call it the Farey division of $[0,1)$ with respect to (h,t).

Let $f(x) = \alpha_k x^k + \cdots + \alpha_1 x$ be a polynomial with real coefficients and let

$$S(f(x)) = S(f(x), T) = \sum_{x=1}^{T} e(f(x)).$$

Since, for integral values of m,

$$\int_0^1 e(\alpha m)\,d\alpha = \begin{cases} 1, & \text{if } m = 0, \\ 0, & \text{if } m \ne 0, \end{cases}$$

we have

$$\begin{aligned} r_s(N) &= \int_0^1 S(\alpha x^k)^s e(-\alpha N)\,d\alpha \\ &= \int_{\mathfrak{M}} S(\alpha x^k)^s e(-\alpha N)\,d\alpha + \int_{\mathfrak{m}} S(\alpha x^k)^s e(-\alpha N)\,d\alpha. \end{aligned} \tag{1.3}$$

We denote these last two integrals by I and J respectively, and we proceed to show that I gives the principal term for $r_s(N)$ while J is only of lower order when $s \ge 2^k + 1$.

By the Hardy-Littlewood method, or the circle method, we mean the method above with the accompanying Farey division.

1.3 Auxiliary Lemmas

In this section we state three inequalities without proofs; see [Hua 3] and [Vaughan 1]. Their generalised forms in algebraic number fields will be given in Chapters 2, 3 and 4.

Lemma 1.1 (Hua). *Let $f(x) = a_k x^k + \cdots + a_1 x$ be a polynomial with integer coefficients, and let q be a positive integer satisfying $(a_k, \ldots, a_1, q) = 1$. Then, for any $\epsilon > 0$,*

$$\sum_{x=1}^{q} e\left(\frac{f(x)}{q}\right) \ll q^{-1+1/k+\epsilon}.$$

Here, and in the following, the constants implicit in the $\ll$ sign may depend on ϵ.

Lemma 1.2 (Weyl's inequality). *Let $f(x) = \alpha_k x^k + \cdots + \alpha_1 x$ be a polynomial with real coefficients, and let α_k, a and q satisfy (1.2) with α_k replacing α. Then, for any $\epsilon > 0$,*

$$S(f(x)) \ll T^{1+\epsilon}\left(\frac{1}{q} + \frac{1}{t} + \frac{h}{T^k}\right)^{2^{1-k}}.$$

Lemma 1.3 (Hua's inequality). *Let $f(x)$ be a polynomial with degree k and with integer coefficients. Then, for $1 \le j \le k$ and $\epsilon > 0$,*

$$\int_0^1 \left|S(\alpha f(x))\right|^{2^j} d\alpha \ll T^{2^j - j + \epsilon}.$$

1.4 Major Arcs

We introduce the following notations:

$$z = \alpha - \frac{a}{q}, \qquad G(a,q) = \frac{1}{q}\sum_{x=1}^{q} e\left(\frac{ax^k}{q}\right),$$

$$I(z,T) = \int_0^T e(y^k z)\, dy, \qquad J(N) = \int_{-\infty}^{\infty} I(z,1)^s e(-zNT^{-k})\, dz,$$

where $1 \le a \le q$, $(a,q) = 1$ and $q \le t$. The expression $J(N)$ is called the *singular integral.*

Lemma 1.4. *Let α be a real number. Then*

$$\int_0^T e(\alpha y^k)\,dy \ll \min\left(T,\,|\alpha|^{-1/k}\right).$$

Proof. Write $w = |\alpha| y^k$. Then, for $\alpha \neq 0$,

$$\int_0^T e(\alpha y^k)\,dy = |\alpha|^{-1/k} k^{-1} \int_0^{|\alpha| T^k} e(\pm w)\, w^{1/k-1}\,dw \ll |\alpha|^{-1/k},$$

and the lemma follows. □

Lemma 1.5. *If $\alpha \in \mathfrak{M}(a,q)$, then*

$$S(\alpha x^k) = G(a,q)\, I(z,T) + O(T^{1-\delta}).$$

Proof. Let b and w be integers satisfying $1 \le b \le q$ and $-1 \le w \le T/q$. Then

$$\int_w^{w+1} e\big((qy+b)^k z\big)\,dy - e\big((qw+b)^k z\big) \ll q|z||y-w|(q|w|+b)^{k-1}$$
$$\ll h^{-1}T^{k-1} \ll T^{-\delta}.$$

Therefore

$$\begin{aligned}\sum_{\substack{1 \le x+b \le T \\ q|x}} e\big((x+b)^k z\big) &= \sum_{\frac{1-b}{q} \le w \le \frac{T-b}{q}} e\big((qw+b)^k z\big) \\ &= \int_{\frac{1-b}{q}}^{\frac{T-b}{q}} e\big((qy+b)^k z\big)\,dy + O(1) + O\big(q^{-1}T^{1-\delta}\big) \\ &= q^{-1} I(z,T) + O\big(q^{-1}T^{1-\delta}\big).\end{aligned}$$

The lemma now follows from

$$S(\alpha x^k) = \sum_{b=1}^{q} e\left(\frac{ab^k}{q}\right) \sum_{\substack{1 \le x+b \le T \\ q|x}} e\big((x+b)^k z\big). \qquad \square$$

Lemma 1.6. *If $s \ge 2k+1$, then*

$$\int_{\mathfrak{M}} S(\alpha x^k)^s e(-\alpha N)\,d\alpha =$$
$$\sum_{q=1}^{t} {\sum_{a=1}^{q}}' G(a,q)^s e\left(-\frac{aN}{q}\right) \int_{-\infty}^{\infty} I(z,T)^s e(-zN)\,dz + O\big(T^{s-k-\delta}\big).$$

Proof. By $\left|S(\alpha x^k)\right| \le T$, $|G(a,q)\, I(z,T)| \le T$ and Lemma 1.5, we have

$$S(\alpha x^k)^s = G(a,q)^s I(z,T)^s + O\big(T^{s-1}|S(\alpha x^k) - G(a,q)I(z,T)|\big)$$
$$= G(a,q)^s I(z,T)^s + O(T^{s-\delta}).$$

Since

$$\sum_{q=1}^{t}\sum_{a=1}^{q}{}' h^{-1}q^{-1}T^{s-\delta} \ll th^{-1}T^{s-\delta} \ll T^{s-k-\delta},$$

we have

$$\int_{\mathfrak{M}} S(\alpha x^k)^s e(-\alpha N)\,d\alpha$$

(1.4)
$$= \sum_{q=1}^{t}\sum_{a=1}^{q}{}' G(a,q)^s e\Big(-\frac{aN}{q}\Big)\int_{-(qh)^{-1}}^{(qh)^{-1}} I(z,T)^s e(-zN)\,dz + O(T^{s-k-\delta}).$$

By Lemmas 1.1 and 1.4 we have, with $\epsilon = 1/2k^2$,

$$\sum_{q=1}^{t}\sum_{a=1}^{q}{}' G(a,q)^s e\Big(-\frac{aN}{q}\Big)\Big(\int_{(qh)^{-1}}^{\infty} I(z,T)^s e(-zN)dz + \int_{-\infty}^{-(qh)^{-1}} I(z,T)^s e(-zN)dz\Big)$$
$$\ll \sum_{q=1}^{t} q^{-s/k+1+\epsilon}\int_{(qh)^{-1}}^{\infty} z^{-s/k}dz \ll \sum_{q=1}^{t}((qh)^{-1})^{-s/k+1}q^{-s/k+1+\epsilon}$$
$$\ll \sum_{q=1}^{t} q^{\epsilon}h^{s/k-1} \ll t^{1+\epsilon}T^{s-k-(1-\delta)(s/k-1)} \ll T^{s-k-\delta}.$$

Substituting this into (1.4) the lemma follows. □

Lemma 1.7. *If $s \geq 2k+1$, then*

$$\int_{\mathfrak{M}} S(\alpha x^k)^s e(-\alpha N)d\alpha = \mathfrak{S}(N)J(N)T^{s-k} + O(T^{s-k-\delta}).$$

Proof. Let $z = T^{-k}z'$ and $y = Ty'$. Replacing z' and y' by z and y, we have $I(z,T) = I(z,1)T$ and

(1.5)
$$\int_{-\infty}^{\infty} I(z,T)^s e(-zN)\,dz = T^{s-k}J(N).$$

Since $s \geq 2k+1$, we have, by Lemma 1.1 with $\epsilon = 1/2k^2$,

$$\sum_{q=t+1}^{\infty}\sum_{a=1}^{q}{}' G(a,q)^s e\Big(-\frac{aN}{q}\Big) \ll \sum_{q=t}^{\infty} q^{-s/k+1+\epsilon} \ll \sum_{q=t}^{\infty} q^{-1-1/k+\epsilon}$$
$$\ll T^{-(1-\delta)(1/k-\epsilon)} \ll T^{-\delta},$$

and so

$$\sum_{q=1}^{t}\sum_{a=1}^{q}{}' G(a,q)^s e\left(-\frac{aN}{q}\right) = \mathfrak{S}(N) + O(T^{-\delta}). \tag{1.6}$$

By Lemma 1.4, we have $J(N) \ll 1$. Therefore the lemma follows by (1.5), (1.6) and Lemma 1.6. □

1.5 Singular Integral

Lemma 1.8. *We have*

$$J(N) = \Gamma\left(1+\frac{1}{k}\right)^s \Gamma\left(\frac{s}{k}\right)^{-1} N^{s/k-1}T^{-s+k}.$$

To prove Lemma 1.8 we shall need:

Lemma 1.9. *Let $F(x)$ be a finite product of bounded monotonic functions over $(0,c]$. Then*

$$\lim_{\lambda\to\infty}\int_0^c F(x)\frac{\sin 2\pi\lambda x}{\pi x}\,dx = \frac{1}{2}F(+0).$$

See, for example, Lemma 6.8.

Lemma 1.10. *If $0<\mu\le 1$, then*

$$\begin{aligned} F_s(\mu) &= \int\limits_{\substack{u_i>0\\ u_1+\cdots+u_{s-1}<\mu}} \left(u_1\cdots u_{s-1}(\mu-u_1-\cdots-u_{s-1})\right)^{1/k-1} du_1\cdots du_{s-1} \\ &= \Gamma\left(\frac{1}{k}\right)^s \Gamma\left(\frac{s}{k}\right)^{-1} \mu^{s/k-1}. \end{aligned}$$

Proof. Let $u_i = \mu v_i\,(1\le i\le s-1)$. Then

$$F_s(\mu) = \mu^{s/k-1}\int\limits_{\substack{v_i>0\\ v_1+\cdots+v_{s-1}<1}} \left(v_1\cdots v_{s-1}(1-v_1-\cdots-v_{s-1})\right)^{1/k-1} dv_1\cdots dv_{s-1}.$$

Let $1-v_1 = w_1$ and $v_i = w_1 w_i\,(2\le i\le s-1)$. The Jacobian of $v_1,\ldots,v_{s-1}$ with respect to $w_1,\ldots,w_{s-1}$ is w_1^{s-2}. We have

$$F_s(\mu) = \mu^{s/k-1} \int_0^1 w_1^{\frac{s-1}{k}-1}(1-w_1)^{1/k-1} dw_1 \times$$

$$\times \int_{\substack{w_i>0 \\ w_2+\cdots+w_{s-1}<1}} \left(w_2 \cdots w_{s-1}(1-w_2-\cdots-w_{s-1})\right)^{1/k-1} dw_2 \cdots dw_{s-1}$$

$$= \mu^{s/k-1}\beta\left(\frac{s-1}{k}, \frac{1}{k}\right) F_{s-1}(1) = \cdots =$$

$$= \mu^{s/k-1}\beta\left(\frac{s-1}{k}, \frac{1}{k}\right)\beta\left(\frac{s-2}{k}, \frac{1}{k}\right)\cdots\beta\left(\frac{1}{k}, \frac{1}{k}\right)$$

$$= \Gamma\left(\frac{1}{k}\right)^s \Gamma\left(\frac{s}{k}\right)^{-1} \mu^{s/k-1}.$$

The lemma is proved. □

Proof of Lemma 1.8. Let $y^k = w$ and $\mu = NT^{-k}$. Then $0 < \mu \le 1$ and

$$k^s J(N) = \int_{-\infty}^{\infty} \left(\int_0^1 w^{1/k-1} e(wz)\, dw\right)^s e(-\mu z)\, dz$$

$$= \lim_{\lambda\to\infty} \int_{0<w_i<1} (w_1 \cdots w_s)^{1/k-1} \frac{\sin 2\pi\lambda(w_1+\cdots+w_s-\mu)}{\pi(w_1+\cdots+w_s-\mu)}\, dw_1 \cdots dw_s.$$

Let $w_1 + \cdots + w_s = u$. Then

$$k^s J(N) = \lim_{\lambda\to\infty} \int_0^s F(u) \frac{\sin 2\pi\lambda(u-\mu)}{\pi(u-\mu)}\, du,$$

where

$$F(u) = \int_{\substack{0<w_i<1 \\ u-1<w_1+\cdots+w_{s-1}<u}} \left(w_1 \cdots w_{s-1}(u-w_1-\cdots-w_{s-1})\right)^{1/k-1} dw_1 \cdots dw_{s-1}.$$

Let $w_i = uv_i\,(1 \le i \le s-1)$. Then

$$F(u) = u^{s/k-1} \int_{\substack{0<v_1<u^{-1} \\ 1-u^{-1}<z_1+\cdots+z_{s-1}<1}} \left(v_1 \cdots v_{s-1}(1-v_1-\cdots-v_{s-1})\right)^{1/k-1} dv_1 \cdots dv_{s-1}$$

is a product of two monotonic functions. Therefore, by Lemmas 1.9 and 1.10, we have

$$k^s J(N) = F(\mu) = F_s(\mu) = \Gamma\left(\frac{1}{k}\right)^s \Gamma\left(\frac{s}{k}\right)^{-1} \mu^{s/k-1},$$

and the lemma is proved. □

1.6 Singular Series

Let p be a prime number and write

$$A(q) = \sum_{a=1}^{q}{}' G(a,q)^s e\left(-\frac{aN}{q}\right), \qquad \chi(p) = \sum_{i=0}^{\infty} A(p^i).$$

We denote by $M(q)$ the number of solutions to the congruence

$$x_1^k + \cdots + x_s^k \equiv N \pmod q,$$

where each x_i runs over a complete residue system $\pmod q$. We also set

$$s_0 = \begin{cases} 2k+1, & \text{if} \quad 2\not| k \quad \text{or} \quad k=2, \\ 4k, & \text{if} \quad 2|k \quad \text{and} \quad k>2, \end{cases}$$

and, for $p^b \| k$,

$$\gamma = \begin{cases} b+1, & \text{if} \quad p>2, \\ b+2, & \text{if} \quad p=2. \end{cases}$$

Lemma 1.11. *The singular series $\mathfrak{S}(N)$ is absolutely convergent if $s \geq 2k+1$. We also have $\mathfrak{S}(N) \geq c_1(k,s) > 0$ if $s \geq s_0$.*

We show that Lemma 1.11 can be derived from the following

Lemma 1.12. *Suppose that $s \geq s_0$. Then $M(p^t) \geq p^{(t-\gamma)(s-1)}$ holds for any prime p and $t \geq \gamma$.*

Lemma 1.13. *If $(q_1,q_2)=1$, then $A(q_1q_2) = A(q_1)A(q_2)$.*

Proof. When each x_i runs over a complete residue system (or a reduced residue system) $\pmod{q_i}$ for $i=1,2$, the expression $x_2q_1 + x_1q_2$ runs over a complete residue system (or a reduced residue system) $\pmod{q_1q_2}$; see, for example, [Hua 1, §2.5]). Therefore

$$\begin{aligned} A(q_1q_2) &= \sum_{x_1=1}^{q_1}{}' \sum_{x_2=1}^{q_2}{}' \left(\frac{1}{q_1q_2} \sum_{y_1=1}^{q_1} \sum_{y_2=1}^{q_2} e\left(\frac{(x_2q_1+x_1q_2)(y_2q_1+y_1q_2)^k}{q_1q_2} \right) \right)^s \\ &\qquad \times e\left(-\frac{(x_2q_1+x_1q_2)N}{q_1q_2} \right) \\ &= \sum_{x_1=1}^{q_1}{}' \left(\frac{1}{q_1} \sum_{y_1=1}^{q_1} e\left(\frac{x_1(q_2y_1)^k}{q_1} \right) \right)^s e\left(-\frac{x_1N}{q_1} \right) \\ &\qquad \times \sum_{x_2=1}^{q_2}{}' \left(\frac{1}{q_2} \sum_{y_2=1}^{q_2} e\left(\frac{x_2(q_1y_2)^k}{q_2} \right) \right)^s e\left(-\frac{x_2N}{q_2} \right) \\ &= A(q_1)A(q_2). \end{aligned}$$

The lemma is proved. □

Lemma 1.14. $M(q) = q^{s-1}\sum_{r|q} A(r)$.

Proof. For any integer m, we have

$$\frac{1}{q}\sum_{a=1}^{q} e\left(\frac{am}{q}\right) = \begin{cases} 1, & \text{if } q|m, \\ 0, & \text{if } q\nmid m, \end{cases}$$

so that

$$\begin{aligned} M(q) &= \frac{1}{q}\sum_{a=1}^{q}\sum_{x_1=1}^{q}\cdots\sum_{x_s=1}^{q} e\left(\frac{a(x_1^k+\cdots+x_s^k-N)}{q}\right) \\ &= q^{s-1}\sum_{r|q}\sum_{b=1}^{r}{}' \left(\frac{1}{r}\sum_{x=1}^{r} e\left(\frac{bx^k}{r}\right)\right)^s e\left(-\frac{bN}{r}\right) = q^{s-1}\sum_{r|q} A(r). \end{aligned}$$

The lemma is proved. □

Lemma 1.15. *If $s \ge 2k+1$, then $\mathfrak{S}(N)$ is absolutely convergent, and*

$$\mathfrak{S}(N) = \prod_p \chi(p).$$

Proof. It follows from Lemma 1.1 that

$$A(q) \ll q^{-s/k+1+\epsilon}.$$

Therefore $\mathfrak{S}(N)$ and $\chi(p)$ are absolutely convergent and

$$\chi(p) = 1 + O(p^{-s/k+1+\epsilon}) \tag{1.7}$$

if $s \ge 2k+1$. Consequently we have, by Lemma 1.13,

$$\prod_{p\le M} \chi(p) = \sum_{q\le M} A(q) + \sum_{q>M}{}'' A(q),$$

where Σ'' means that q runs over all integers with no prime divisor exceeding M. Since $\sum''_{q>M} A(q) = o(1)$ as $M \to \infty$, we have $\prod_p \chi(p) = \mathfrak{S}(N)$, and the lemma is proved. □

Proof of Lemma 1.11. By Lemma 1.15, it suffices to show that $\mathfrak{S}(N) \ge c_1(k,s) > 0$ if $s \ge s_0$. By (1.7) there exists $c_2(k)$ such that

$$\frac{1}{2} \le \prod_{p>c_2} \chi(p) \le \frac{3}{2}.$$

Then, by Lemmas 1.12 and 1.14, we have

$$\chi(p) = \sum_{i=0}^{\infty} A(p^i) = \lim_{t\to\infty} p^{-t(s-1)} M(p^t) \ge p^{-\gamma(s-1)} = c_3(k,p,s),$$

say. Therefore, by Lemma 1.15,

$$\mathfrak{S}(N) = \prod_{p \le c_2} \chi(p) \prod_{p > c_2} \chi(p) \ge \frac{1}{2} \prod_{p \le c_2} c_3 = c_1(k, s) > 0.$$

The lemma is proved. □

1.7 Proof of Lemma 1.12

Lemma 1.16. *Suppose that $p \nmid a$. If the congruence $y^k \equiv a \pmod{p^\gamma}$ is soluble, then so is the congruence $x^k \equiv a \pmod{p^t}$ for $t > \gamma$.*

Proof. 1) Suppose that $p > 2$. Let g be a primitive root mod p^t. Then g is also a primitive root mod p^γ. We have $z \equiv g^{\mathrm{ind}\, z} \pmod{p^t}$ for any $p \nmid z$, and the congruence $y^k \equiv a \pmod{p^\gamma}$ is equivalent to $k \,\mathrm{ind}\, y \equiv \mathrm{ind}\, a \pmod{p^{\gamma-1}(p-1)}$. It follows that $(k, p^{\gamma-1}(p-1)) | \mathrm{ind}\, a$. Since $p^{\gamma-1} \| k$, we see that $k \,\mathrm{ind}\, x \equiv \mathrm{ind}\, a \pmod{p^{t-1}(p-1)}$ is soluble, that is, $x^k \equiv a \pmod{p^t}$ is soluble.

2) Suppose that $p = 2$. If $b = 0$ (that is, if k is odd), then the congruence $x^k \equiv a \pmod{2^t}$ is soluble for any odd a. In fact, if x runs over a reduced residue system mod 2^t, then so does x^k. Now suppose that $b \ge 1$. Since k is even, we have $x^k \equiv 1 \pmod 4$ for all odd x. Those residue classes mod 2^t that are $\equiv 1 \pmod 4$ form a cyclic group of order 2^{t-2} generated by 5; see, for example, [Hua 1; §3.9]. As before, if we use 5 instead of g, the hypothesis becomes $k \,\mathrm{ind}\, y \equiv \mathrm{ind}\, a \pmod{2^{\gamma-2}}$. Since $2^{\gamma-2} \| k$, the lemma follows similarly to the case $p > 2$. □

Lemma 1.17. *If $s \ge s_0$, then the congruence*

$$x_1^k + \cdots + x_s^k \equiv N \pmod{p^\gamma} \tag{1.8}$$

has a solution with at least one x_i not divisible by p.

Proof. If $N \not\equiv 0 \pmod p$, then any solution of (1.8) must contain at least one x_i not divisible by p. If $N \equiv 0 \pmod p$, it suffices to solve

$$x_1^k + \cdots + x_{s-1}^k + 1 \equiv N \pmod{p^\gamma}.$$

Hence it suffices to establish the solubility of (1.8) when $N \not\equiv 0 \pmod p$ for $s \ge s_0 - 1$.

1) Suppose that $p > 2$. Consider all those N with $1 \le N < p^\gamma$ and $p \nmid N$. Let $s(N)$ be the least s for which (1.8) is soluble. If $N \equiv z^k N' \pmod{p^\gamma}$, then $s(N) = s(N')$. Hence if we distribute the numbers N into classes according to $s(N)$, the number in each class is at least equal to the number of distinct values assumed by z^k when $z \not\equiv 0 \pmod p$. The congruence $z^k \equiv a \pmod{p^\gamma}$

is soluble if and only if $p^b(k,p-1) | \operatorname{ind} a$, and therefore the number of distinct values for $a \pmod{p^\gamma}$ is

$$\frac{p^b(p-1)}{p^b(k,p-1)} = \frac{(p-1)}{(k,p-1)} = r,$$

say. This means that each class of values of N includes at least r elements.

Let A_i be the set of N with $s(N) = i$, and let m be the largest integer such that A_m is not empty. Consider the least N' which is not in $A_1, \ldots, A_{j-1}$. Then either $N' - 1$ or $N' - 2$ is not divisible by p, and, being less than N', it must be one of $A_1, \ldots, A_{j-1}$. Representing N' as

$$N' = (N'-1) + 1^k \qquad \text{or} \qquad N' = (N'-2) + 1^k + 1^k,$$

we deduce that $s(N') \leq j+1$. Hence A_j and A_{j+1} cannot be both empty. Since A_1 is not empty, there are at least $(m-1)/2$ sets among $A_1, \ldots, A_{m-1}$ that are not empty, and also A_m is not empty. Since each such set contains at least r numbers, we have

$$\frac{1}{2}(m+1)r \leq \phi(p^\gamma) = p^{\gamma-1}(p-1)$$

so that

$$m+1 \leq \frac{2p^{\gamma-1}(p-1)}{r} = 2p^{\gamma-1}(k,p-1) \leq 2k.$$

Hence $s(N) \leq m \leq 2k-1$ for all N.

2) Suppose that $p = 2$. If $b = 0$ (that is, if k is odd), then $s = 1$ is enough; see the proof of Lemma 1.16. Now suppose that $b \geq 1$. If $k = 2$, then $N = 1, 3, 5, 7$ and so (1.8) has a solution with $x_1 = 1$ when $s \geq 4$. If $k > 2$, then taking x_i in (1.8) to be 0 or 1, we can solve the congruence (1.8) if $s \geq 2^\gamma - 1 = 2^{b+2} - 1 \leq 4k - 1$. The lemma is proved. □

Proof of Lemma 1.12. Suppose that $N \equiv a_1^k + \cdots + a_s^k \pmod{p^\gamma}$, where $0 \leq a_i < p^\gamma \, (1 \leq i \leq s)$ and $p \nmid a_1$. For $t > \gamma$ we choose

$$x_i = a_i + m_i p^\gamma, \qquad 0 \leq m_i < p^{t-\gamma}, \quad 2 \leq i \leq s.$$

Then

$$x_i \equiv a_i \pmod{p^\gamma}, \qquad 0 \leq x_i < p^t, \quad 2 \leq i \leq s.$$

For each $x_2, \ldots, x_s$, we can choose x_1 by Lemma 1.16 such that

$$x_1^k \equiv N - x_2^k - \cdots - x_s^k \pmod{p^t}.$$

Therefore

$$M(p^t) \geq p^{(t-\gamma)(s-1)}.$$

The lemma is proved. □

1.8 Proof of Theorem 1.1

If α satisfies (1.2) and $\alpha \in \mathfrak{m}$, then $q > t$ and, by Lemma 1.2, we have

$$S(\alpha x^k) \ll T^{1-(1-\delta)2^{1-k}+\epsilon/2s}.$$

Then, by Lemma 1.3, for $s \geq 2^k + 1$,

$$\begin{aligned} \int_{\mathfrak{m}} |S(\alpha x^k)|^s d\alpha &\ll T^{(1-(1-\delta)2^{1-k})(s-2^k)+\epsilon/2} \int_0^1 |S(\alpha x^k)|^{2^k} d\alpha \\ &\ll T^{(1-(1-\delta)2^{1-k})(s-2^k)+2^k-k+\epsilon} \ll T^{s-k-\epsilon}, \end{aligned} \tag{1.9}$$

where $\epsilon = (1-\delta)/2^k$. Since $2^k + 1 \geq s_0$, the theorem follows by Lemmas 1.7, 1.8, 1.11 and (1.3) together with (1.9). □

Remark. Let $f_i(x) = a_{ki}x^k + \cdots + a_{1i}x \,(1 \leq i \leq s)$ be given polynomials with integer coefficients and $a_{ki} > 0 \,(1 \leq i \leq s)$. Consider

$$N = f_1(x_1) + \cdots + f_s(x_s). \tag{1.10}$$

Let $R_s(N)$ denote the number of solutions of (1.10), where x_i are positive integers. Besides the lower estimation of the singular series we may prove that if $s \geq 2^k + 1$, then

$$R_s(N) = \mathfrak{S}(N)(a_{k1} \cdots a_{ks})^{-1/k} \Gamma\left(1 + \frac{1}{k}\right)^s \Gamma\left(\frac{s}{k}\right)^{-1} N^{s/k-1}(1 + o(1)),$$

where

$$\mathfrak{S}(N) = \sum_{q=1}^{\infty} {\sum_{a=1}^{q}}' \prod_{s=1}^{s} \left(\frac{1}{q} e\left(\frac{af_i(x)}{q}\right)\right) e\left(-\frac{aN}{q}\right).$$

In fact, we may express $R_s(N)$ by an integral over $[0, 1)$ and divide the interval $[0, 1)$ into $\mathfrak{M}$ and $\mathfrak{m}$ as before. The integral over $\mathfrak{m}$ can be estimated similarly to that in Waring's problem. Using Vinogradov's lemma ([Vinogradov 4; p.199]):

$$\int_0^T e(\beta_k y^k + \cdots + \beta_1 y)\, dy \ll \min\left(T, |\beta_i|^{-1/k} T^{1-i/k}\right), \quad 1 \leq i \leq k, \tag{1.11}$$

instead of Lemma 1.4, we may establish a lemma similar to Lemma 1.7. So we can treat the integral over $\mathfrak{M}$. Using (1.11) again, we have

$$\begin{aligned} &\int_{-\infty}^{\infty} \left\{ \prod_{i=1}^{s} \left(\int_0^T e(f_i(y)x)\, dy \right) - \prod_{i=1}^{s} \left(\int_0^T e(a_{ki}y^k x)\, dy \right) \right\} e(-Nx)\, dx \\ &\ll \int_0^{T^{-k+1/2}} T^{k-1} x T^{s-1} dx + \int_{T^{-k+1/2}}^{\infty} x^{-s/k} dx \ll T^{s-k-\delta}, \end{aligned}$$

and so the singular integral is equal to

$$\int_{-\infty}^{\infty} \prod_{i=1}^{s} \left(\int_0^1 e\big(a_{ki} y^k x\big)\, dy \right) e\big(- N T^{-k} x\big)\, dx$$

which can be treated as before.

Notes

For Theorems 1.1 and 1.2, see [Hua 4], and also [Hua 3], [Davenport 1] and [Vaughan 1]. On the improvements of these theorems the reader should consult [Vinogradov 3], [Hua 4], [Vaughan 1] and the recent papers of Heath-Brown [3], Karatsuba [1], Vaughan [1,3] and Wooley [1].

For additive problems with prime variables and polynomial summands the reader should consult Hua's monographs [Hua 2] and [Hua 4].

Chapter 2

Complete Exponential Sums

2.1 Introduction

Let

$$f(\lambda) = \alpha_k \lambda^k + \cdots + \alpha_1 \lambda$$

be a k-th degree polynomial with coefficients in $\mathbb{K}$. Let $\mathfrak{a} = (\alpha_k, \ldots, \alpha_1)$ be the fractional ideal generated by $(\alpha_k, \ldots, \alpha_1)$. Suppose that $\mathfrak{a}\delta = \mathfrak{g}/\mathfrak{q}$, where $\mathfrak{g}$, $\mathfrak{q}$ are two relatively prime ideals, and

$$S(f(x), \mathfrak{q}) = \sum_{\lambda(\mathfrak{q})} E(f(\lambda)),$$

where λ runs over a complete residue system mod $\mathfrak{q}$.

Note that the definition of the sum $S(f(x), \mathfrak{q})$ is independent of the choice of the residue system mod $\mathfrak{q}$. In fact the definition of δ can be stated in the following way: δ^{-1} is the aggregate of all numbers ξ of $\mathbb{K}$ such that $E(\xi, \alpha) = 1$ for all α of J. Consequently if $\beta \in (\mathfrak{q}\delta)^{-1}$ and $\alpha_1 \equiv \alpha_2 \pmod{\mathfrak{q}}$, then $E(\beta\alpha_1) = E(\beta\alpha_2)$, and the assertion follows. The sum $S(f(x), \mathfrak{q})$ is called the *complete exponential sum* of $f(\lambda)$ (mod $\mathfrak{q}$). The aim of this chapter is to prove the following

Theorem 2.1 (Hua). *We have, for any $\epsilon > 0$,*

$$S(f(x), \mathfrak{q}) = O(N(\mathfrak{q})^{1-1/k+\epsilon}),$$

where the constant in the symbol O depends only on k, n and ϵ.

For the case of the prime ideal $\wp$ and $f(\lambda) = \alpha\lambda^k$, where $\alpha\delta = \mathfrak{g}/\wp$ with $\wp \nmid \mathfrak{g}$, we have

Theorem 2.2. *Let $d = (k, N(\wp) - 1)$. Then*

$$|S(\alpha\lambda^k, \wp)| \le (d-1)N(\wp)^{1/2}.$$

2.2 Several Lemmas

Let

$$\mathfrak{q} = \wp_1^{\ell_1} \cdots \wp_s^{\ell_s},$$

where the $\wp_i$ are distinct prime ideals. The number of divisors of $\mathfrak{q}$ is given by the divisor function

$$d(\mathfrak{q}) = \prod_{i=1}^{s} (\ell_i + 1).$$

Lemma 2.1. *We have*

$$d(\mathfrak{q}) = O(N(\mathfrak{q})^{\epsilon}).$$

Proof. Since $N(\wp_i) = p_i^{a_i}$, where p_i is a rational prime factor of $N(\mathfrak{q})$ and $1 \leq a_i \leq n$, we have

$$d(\mathfrak{q}) \leq d(N(\mathfrak{q})),$$

where $d(N(\mathfrak{q}))$ is the ordinary divisor function. The lemma follows; see, for example, [Hua 1; §6.5]. □

Lemma 2.2. *Let $\alpha_m, \ldots, \alpha_0$ be integers of $\mathbb{K}$ and $\mathfrak{a} = (\alpha_m, \ldots, \alpha_0)$. If $\wp \nmid \mathfrak{a}$, then the number of incongruent solutions* (mod $\wp$) *(including repeated ones) to the congruence*

$$f(\lambda) = \alpha_m \lambda^m + \cdots + \alpha_0 \pmod{\wp}$$

does not exceed m.

The proof is similar to that of the rational case; see, for example, [Hua 1; §2.9] and [Hecke 1; §27].

Note that if $\mathfrak{a}$ is a fractional ideal and $\mathfrak{a}\delta = \mathfrak{g}/\wp^{\ell}$, $(\mathfrak{g}, \wp) = 1$, the result holds for the congruence

$$f(\lambda) \equiv 0 \pmod{\wp^{-\ell+1}}.$$

Lemma 2.3. *Let $f(\lambda)$ be a polynomial with coefficients in J and α be a root of multiplicity m of the congruence*

$$f(\lambda) \equiv 0 \pmod{\wp}.$$

Let π be an integer such that $\wp \| \pi$ and let u the rational integer satisfying $\wp^u \| \big(f(\pi\lambda + \alpha) - f(\alpha)\big)$. Further, let

$$g(\lambda) = \pi^{-u}\big(f(\pi\lambda + \alpha) - f(\alpha)\big) \pmod{\wp}$$

be a polynomial with coefficients in J. Then $u \leq m$ and the congruence

$$g(\lambda) \equiv 0 \pmod{\wp}$$

has at most m solutions.

Proof. Without loss of generality we may assume that $\alpha = 0$. Then

$$f(\lambda) = \lambda^m f_1(\lambda) + f_2(\lambda), \qquad f_1(0) \not\equiv 0 \pmod{\wp},$$

where $f_2(\lambda)$ is a polynomial of degree less than m and all its coefficients are divisible by $\wp$. Now we have

$$f(\pi\lambda) = \pi^m \lambda^m f_1(\pi\lambda) + f_2(\pi\lambda).$$

Since the coefficient of λ^m is equal to $\pi^m f_1(0)$ which is not divisible by $\wp^{m+1}$, we have $u \leq m$. Since $\pi^{-u} f(\pi\lambda)$ is congruent to a polynomial of degree at most $m \pmod{\wp}$, the lemma follows. □

Lemma 2.4. *If* $(\mathfrak{q}_1, \mathfrak{q}_2) = 1$ *and* $f(0) = 0$, *then there are two polynomials* $f_1(\lambda)$ *and* $f_2(\lambda)$, *each of degree* k, *such that*

$$S(f(\lambda), \mathfrak{q}_1\mathfrak{q}_2) = S(f_1(\lambda), \mathfrak{q}_1) S(f_2(\lambda), \mathfrak{q}_2).$$

Proof. We can find two integers λ_1 and λ_2 such that

$$(\lambda_1, \mathfrak{q}_1\mathfrak{q}_2) = \mathfrak{q}_1 \qquad \text{and} \qquad (\lambda_2, \mathfrak{q}_1\mathfrak{q}_2) = \mathfrak{q}_2;$$

see, for example, [Hecke 1; §26]. Put

$$\lambda = \lambda_1\mu_2 + \lambda_2\mu_1.$$

Then if μ_1 and μ_2 run over complete residue systems $\pmod{\mathfrak{q}_1}$ and $\pmod{\mathfrak{q}_2}$ respectively, λ runs over a complete residue system $\pmod{\mathfrak{q}_1\mathfrak{q}_2}$. This can be proved similarly to that of the rational case; see, for example, [Hua 1; §2.5]. We have

$$\begin{aligned} S(f(\lambda), \mathfrak{q}_1\mathfrak{q}_2) &= \sum_{\mu_1(\mathfrak{q}_1)} \sum_{\mu_2(\mathfrak{q}_2)} E(f(\lambda_1\mu_2 + \lambda_2\mu_1)) \\ &= \sum_{\mu_1(\mathfrak{q}_1)} E(f(\lambda_2\mu_1)) \sum_{\mu_2(\mathfrak{q}_2)} E(f(\lambda_1\mu_2)) \\ &= S(f_1(\lambda), \mathfrak{q}_1) S(f_2(\lambda), \mathfrak{q}_2), \end{aligned}$$

where $f_1(\lambda) = f(\lambda_2\lambda)$ and $f_2(\lambda) = f(\lambda_1\lambda)$. Now we have to verify that the ideal generated by the coefficients of $f_1(\lambda)$ can be expressed as $\mathfrak{g}(\delta\mathfrak{q}_1)^{-1}$, where $(\mathfrak{g}, \mathfrak{q}_1) = 1$, but this is quite evident. The lemma is proved. □

Lemma 2.5. *Let* $\mathfrak{q}$ *be an integral ideal and* α *be an integer. Then*

$$\sum_{\xi} E(\xi\alpha) = \begin{cases} N(\mathfrak{q}), & \text{if } \mathfrak{q} | \alpha, \\ 0, & \text{if } \mathfrak{q} \nmid \alpha, \end{cases}$$

where ξ *runs over a complete residue system of* $(\mathfrak{q}\delta)^{-1} \bmod \delta^{-1}$.

Proof. First we note that the sum in the lemma is independent of the choice of the residue system of $(\mathfrak{q}\delta)^{-1} \bmod \delta^{-1}$. If $\mathfrak{q}|\alpha$, then $\xi\alpha \in \delta^{-1}$ and $E(\xi\alpha) = 1$ for all ξ. We have the first conclusion.

If $\mathfrak{q}\not|\alpha$, there is an element ξ_0, which belongs to $(\mathfrak{q}\delta)^{-1}$, but $\xi_0\alpha$ does not belong to δ^{-1}. In fact, if for all ξ_0 belonging to $(\mathfrak{q}\delta)^{-1}$ we have $\xi_0\alpha$ belonging to δ^{-1}, then

$$\delta^{-1}|(\mathfrak{q}\delta)^{-1}\alpha.$$

Consequently $\mathfrak{q}|\alpha$. This is impossible. By the definition of δ^{-1}, there is an integer γ such that

$$E(\gamma\xi_0\alpha) \neq 1.$$

Since $\gamma\xi_0$ belongs to $(\mathfrak{q}\delta)^{-1}$, we obtain

$$\sum_{\xi} E(\xi\alpha) = \sum_{\xi} E((\xi + \gamma\xi_0)\alpha) = E(\gamma\xi_0\alpha) \sum_{\xi} E(\xi\alpha).$$

The second conclusion of the lemma follows. □

2.3 Mordell's Lemma

For prime ideal $\wp$, $S(f(\lambda), \wp)$ can be estimated by Mordell's method.

Lemma 2.6 (Mordell). *We have*

$$|S(f(\lambda), \wp)| \leq k^n N(\wp)^{1-1/k}.$$

Proof. Without loss of generality, we may assume that α_k does not belong to δ^{-1}, for otherwise

$$S(f(\lambda), \wp) = S(f(\lambda) - \alpha_k\lambda^k, \wp),$$

since $E(\alpha_k\lambda^k) = 1$ for any integer λ. This means that $f(\lambda)$ can be replaced by a polynomial of lower order. Thus we now assume that α_k belongs to $(\wp\delta)^{-1}$ but not to δ^{-1}. The lemma is trivial if $N(\wp) \leq k^n$, since

$$|S(f(\lambda), \wp)| \leq N(\wp) \leq k^n N(\wp)^{1-1/k}.$$

Now assume $N(\wp) > k^n$ and consequently $\wp\not|p$ if $1 \leq p \leq k$. We have

$$|S(f(\lambda), \wp)|^{2k} = \frac{1}{N(\wp)(N(\wp) - 1)} \sum_{\nu(\wp)}{}' \sum_{\mu(\wp)} |S(f(\nu\lambda + \mu), \wp)|^{2k},$$

where ν runs over a reduced residue system mod $\wp$. Write

$$f(\nu\lambda + \mu) = \beta_k\lambda^k + \cdots + \beta_1,$$

where

$$\beta_k \equiv \alpha_k \nu^k \pmod{\delta^{-1}},$$
$$\beta_{k-1} \equiv k\alpha_k \nu^{k-1}\mu + \alpha_{k-1}\nu^{k-1} \pmod{\delta^{-1}},$$

and so on. For fixed $\beta_k, \beta_{k-1}, \ldots$ belonging to $(\wp\delta)^{-1}$, the number of integers $\nu\,(\notin \wp)$ does not exceed k. In fact, there is an integer $\tau \in \wp\delta$ but $\tau\alpha_k \notin \wp$. Otherwise if $\tau\alpha_k \in \wp$ for all $\tau \in \wp\delta$, we have $\wp | \wp\delta\alpha_k$, i.e., $\delta^{-1} | \alpha_k$ which is impossible. Consequently $\tau\alpha_k\nu^k \equiv \tau\beta_k \pmod{\wp}$ has at most k solutions by Lemma 1.2. For a fixed $\nu\,(\wp \nmid \nu)$, the congruence $\tau k\alpha_k\nu^{k-1}\mu + \tau\alpha_{k-1}\nu^{k-1} \equiv \tau\beta_{k-1} \pmod{\wp}$ has a unique solution μ since $\wp \nmid \tau k \alpha_k \nu^{k-1}$. Therefore we have

$$|S(f(\lambda), \wp)|^{2k} \le \frac{k}{N(\wp)(N(\wp)-1)} \sum_{\beta_k} \cdots \sum_{\beta_1} |S(\beta_k\lambda^k + \cdots + \beta_1\lambda, \wp)|^{2k},$$

where each β_i runs over a complete residue system of $(\wp\delta)^{-1} \bmod \delta^{-1}$. By Lemma 2.5, we have

$$\sum_{\beta_k} \cdots \sum_{\beta_1} |S(\beta_k\lambda^k + \cdots + \beta_1\lambda, \wp)|^{2k}$$
$$= \sum_{\beta_k} \cdots \sum_{\beta_1} \sum_{\lambda_1(\wp)} \cdots \sum_{\lambda_k(\wp)} \sum_{\mu_1(\wp)} \cdots \sum_{\mu_k(\wp)} E(\psi) = N(\wp)^k M,$$

where

$$\psi = \beta_k(\lambda_1^k + \cdots + \lambda_k^k - \mu_1^k - \cdots - \mu_k^k) + \cdots + \beta_1(\lambda_1 + \cdots + \lambda_k - \mu_1 - \cdots - \mu_k)$$

and M is equal to the number of solutions of the system of congruences

$$\lambda_1^h + \cdots + \lambda_k^h \equiv \mu_1^h + \cdots + \mu_k^h \pmod{\wp}, \qquad 1 \le h \le k.$$

By a theorem on symmetric functions, we deduce immediately

$$(X-\lambda_1)\cdots(X-\lambda_k) \equiv (X-\mu_1)\cdots(X-\mu_k) \pmod{\wp},$$

since $\wp \nmid k!$. Then we have that $\lambda_1, \ldots, \lambda_k$ are a permutation of $\mu_1, \ldots, \mu_k$ and thus

$$M \le k! N(\wp)^k.$$

Consequently we have

$$|S(f(\lambda), \wp)|^{2k} \le \frac{k \cdot k!}{N(\wp)(N(\wp)-1)} N(\wp)^{2k}$$
$$\le 2k \cdot k!\, N(\wp)^{2k-2} \le k^{2k} N(\wp)^{2k-2},$$

and the lemma is proved. □

2.4 Fundamental Lemma

Lemma 2.7. *We have*

$$|S(f(\lambda),\wp^\ell)| \le k^{2n+1}N(\wp^\ell)^{1-1/k}. \tag{2.1}$$

Proof. Let $\mathfrak{b} = (k\alpha_k, \ldots, \alpha_1)$. Then $\mathfrak{a}|\mathfrak{b}$. Suppose that $\wp^t \| \mathfrak{b}\mathfrak{a}^{-1}$. Let m be the number of solutions, counted according to multipicity, of the congruence

$$f'(\lambda) \equiv 0 \pmod{\wp^{-\ell+t+1}} \tag{2.2}$$

as λ runs over a complete residue system $\pmod{\wp}$. We have $m \le k-1$ by Lemma 2.2. Evidently, (2.1) is a consequence of the sharper result

$$|S(f(\lambda),\wp^\ell)| \le k^{2n}\max(1,m)N(\wp^\ell)^{1-1/k}. \tag{2.3}$$

If $t \ge 1$, then $\wp^t$ divides at least one of the integers $k, \ldots, 2$ and so $N(\wp^t) \le k^n$, that is

$$N(\wp) \le k^{n/t}. \tag{2.4}$$

Suppose that $\ell < 2(t+1)$. If $t = 0$, we have $\ell = 1$ and (2.3) follows from Lemma 2.6. If $t \ge 1$, then by (2.4)

$$\begin{aligned} |S(f(\lambda),\wp^\ell)| &\le N(\wp)^\ell \le N(\wp)^{\ell(1-1/k)}N(\wp)^{\frac{2t+1}{k}} \\ &\le N(\wp)^{\ell(1-1/k)}k^{(2+1/t)n/k} \le k^{2n}N(\wp)^{\ell(1-1/k)}; \end{aligned}$$

in other words, (2.3) is true for $\ell < 2(t+1)$. Now assume that $\ell \ge 2(t+1)$ and (2.3) is true for smaller ℓ. Let $\mu_1, \ldots, \mu_r$ be the distinct roots of (2.2) with multiplicities $m_1, \ldots, m_r$ respectively. Then $m_1 + \cdots + m_r = m$. Evidently

$$S(f(\lambda),\wp^\ell) = \sum_{\nu(\wp)} \sum_{\substack{\lambda(\wp^\ell) \\ \lambda\equiv\nu \pmod{\wp}}} E(f(\lambda)) = \sum_{\nu(\wp)} S_\nu,$$

say. If ν is not one of the μ's, then letting

$$\lambda = \tau + \pi^{\ell-t-1}\omega, \qquad \wp\|\pi,$$

we have

$$\begin{aligned} S_\nu &= \sum_{\substack{\tau(\wp^{\ell-t-1}) \\ \tau\equiv\nu \pmod{\wp}}} \sum_{\omega(\wp^{t+1})} E(f(\tau) + \pi^{\ell-t-1}\omega f'(\tau)) \\ &= \sum_{\substack{\tau(\wp^{\ell-t-1}) \\ \tau\equiv\nu \pmod{\wp}}} E(f(\tau)) \sum_{\omega(\wp^{t+1})} E(\pi^{\ell-t-1}\omega f'(\tau)) = 0 \end{aligned}$$

by Lemma 2.5, since $\wp \nmid \pi^{\ell-t}f'(\tau)$. Therefore

$$\begin{aligned}|S(f(\lambda),\wp^\ell)| &\le \sum_{s=1}^{r} \Big| \sum_{\lambda(\wp^{\ell-1})} E(f(\mu_s+\pi\lambda))\Big| \\ &= \sum_{s=1}^{r} \Big| \sum_{\lambda(\wp^{\ell-1})} E(f(\mu_s+\pi\lambda)-f(\mu_s))\Big|.\end{aligned}$$

Let $\mathfrak{L}$ be the ideal generated by the coefficients of

$$f_s(\lambda) = f(\mu_s+\pi\lambda) - f(\mu_s).$$

Then $\mathfrak{a}|\mathfrak{L}$. We define σ_s by $\wp^{\sigma_s}\|\mathfrak{L}\mathfrak{a}^{-1}$. Also, if $\ell \le \sigma_s$, we use the conventional meaning

$$S(f_s(\lambda),\wp^{\ell-\sigma_s}) = N(\wp)^{\ell-\sigma_s}.$$

Then we have

$$|S(f(\lambda),\wp^\ell)| \le \sum_{s=1}^{r} N(\wp)^{\sigma_s-1}|S(f_s(\lambda),\wp^{\ell-\sigma_s})|. \tag{2.5}$$

Now we are going to prove that

$$1 \le \sigma_s \le k. \tag{2.6}$$

If (2.6) is not true, then $\wp^{-\ell+k+1}$ divides all the coefficients of $f_s(\lambda)$, that is

$$\wp^{-\ell+k+1}\Big|\frac{f^{(r)}(\mu_s)}{r!}\pi^r, \qquad 1\le r\le k.$$

Consequently

$$\wp^{-\ell+1}\Big|\frac{f^{(r)}(\mu_s)}{r!}, \qquad 1\le r\le k.$$

Since $f^{(r)}(\mu_s)/r!$ is equal to α_r plus a linear combination of $\alpha_k,\dots,\alpha_{r+1}$ with integral coefficients, we deduce successively that $\wp^{-\ell+1}|\alpha_k,\dots,\wp^{-\ell+1}|\alpha_1$. This leads to a contradiction with $\mathfrak{q}=\wp^\ell$.

Let $\mathfrak{L}'$ be the ideal generated by the coefficients of $f_s'(\lambda)$ and $\wp^u\|\mathfrak{L}'\mathfrak{L}^{-1}$. Then by Lemma 2.3, the number of solutions of the congruence $f_s'(\lambda)\equiv 0 \pmod{\wp^{u+1}}$ does not exceed m_s. By (2.5), (2.6) and the induction hypothesis, we have, for $\ell > \max(\sigma_1,\dots,\sigma_s)$,

$$|S(f(\lambda),\wp^\ell)| \le k^{2n}\sum_{s=1}^{r} N(\wp)^{\sigma_s(1-1/k)} m_s N(\wp)^{(\ell-\sigma_k)((1-1/k)} = k^{2n} m N(\wp)^{\ell(1-1/k)}.$$

For the case $\ell\le\max(\sigma_1,\dots,\sigma_s)$, we have $\ell\le k$ by (2.6), and by (2.5),

$$|S(f(\lambda),\wp^\ell)| \le rN(\wp)^{\ell-1} \le mN(\wp)^{\ell(1-1/k)}.$$

Therefore we have (2.3) and consequently (2.1). Note that if $\sum_{s=1}^{r} m_s = 0$, the method shows that $S(f(\lambda),\wp^\ell)=0$ if $\ell\ge 2(t+1)$. The lemma is proved. □

2.5 Proof of Theorem 2.1.

Let

$$\mathfrak{q} = \wp_1^{\ell_1} \cdots \wp_s^{\ell_s},$$

where $\wp_i (1 \le i \le s)$ are distinct prime ideals. Then we have, by repeated applications of Lemma 2.4,

$$S(f(\lambda), \mathfrak{q}) = \prod_{i=1}^{s} S(f_i(\lambda), \wp_i^{\ell_i}).$$

By Lemmas 2.1 and 2.7 we have

$$\begin{aligned} |S(f(\lambda), \mathfrak{q})| &\le \prod_{i=1}^{s} k^{2n+1} N(\wp_i^{\ell_i})^{1-1/k} \\ &\le \prod_{i=1}^{s} (1+\ell_i)^{(2n+1)\log k/\log 2} N(\wp_i^{\ell_i})^{1-1/k} \\ &= d(\mathfrak{q})^{(2n+1)\log k/\log 2} N(\mathfrak{q})^{1-1/k} \ll N(\mathfrak{q})^{1-1/k+\epsilon}. \end{aligned}$$

The theorem is proved. □

2.6 Proof of Theorem 2.2.

Let η be a primitive root $\pmod{\wp}$. Then $\mu \equiv \eta^{\operatorname{ind}\mu} \pmod{\wp}$ for any μ with $\wp \nmid \mu$. The necessary and sufficient condition for the solubility of the congruence $\lambda^k \equiv \mu \pmod{\wp}$ is $k \operatorname{ind}\lambda \equiv \operatorname{ind}\mu \pmod{N(\wp)-1}$ or $d | \operatorname{ind}\mu$. If it is soluble, it has d incongruent solutions $\pmod{\wp}$.

Suppose that $d = 1$. If μ runs over a complete residue system $\pmod{\wp}$, then $\alpha\mu$ runs over a complete residue system of $(\wp\delta)^{-1}$, $\pmod{\delta^{-1}}$. Therefore by Lemma 2.5 we have

$$S(\alpha\lambda^k, \wp) = \sum_{\mu(\wp)} E(\alpha\mu) = 0.$$

Suppose now that $d > 1$. Then

$$S(\alpha\lambda^k, \wp) = 1 + \sum_{\mu(\wp)}{}' E(\alpha\mu) \sum_{m=0}^{d-1} e\left(m\frac{\operatorname{ind}\mu}{d}\right).$$

Since $\sum_{\mu(\wp)}{}' E(\alpha\mu) = -1$, we have

$$S(\alpha\lambda^k, \wp) = \sum_{m=1}^{d-1} \sum_{\mu(\wp)}{}' e\left(m\frac{\operatorname{ind}\mu}{d}\right) E(\alpha\mu),$$

and it follows by Schwarz's inequality that

$$\left|S(\alpha\lambda^k,\wp)\right|^2 \le (d-1)\sum_{m=1}^{d-1}\sum_{\mu(\wp)}{}'\sum_{\mu_1(\wp)}{}' e\left(m\frac{\operatorname{ind}\mu-\operatorname{ind}\mu_1}{d}\right)E(\alpha(\mu-\mu_1)).$$

For any given μ and μ_1, let τ be the number such that

$$\mu \equiv \tau\mu_1 \pmod{\wp}.$$

Then

$$\begin{aligned}
\left|S(\alpha\lambda^k,\wp)\right|^2 &\le (d-1)\sum_{m=1}^{d-1}\sum_{\tau(\wp)}{}'\sum_{\mu_1(\wp)}{}' e\left(m\frac{\operatorname{ind}\tau}{d}\right)E(\alpha(\tau-1)\mu_1)\\
&= (d-1)\sum_{m=1}^{d-1}\left(N(\wp)-1+\sum_{\tau(\wp)}{}'' e\left(m\frac{\operatorname{ind}\tau}{d}\right)\sum_{\mu_1(\wp)}{}' E(\alpha(\tau-1)\mu_1)\right)\\
&= (d-1)\sum_{m=1}^{d-1}\left(N(\wp)-1-\sum_{\tau(\wp)}{}'' e\left(m\frac{\operatorname{ind}\tau}{d}\right)\right),
\end{aligned}$$

where Σ'' denotes a sum with the term $\tau=1$ deleted. Since

$$\begin{aligned}
\sum_{\tau(\wp)}{}'' e\left(m\frac{\operatorname{ind}\tau}{d}\right) &= -1+\sum_{\tau(\wp)}{}' e\left(m\frac{\operatorname{ind}\tau}{d}\right)\\
&= -1+\frac{N\wp-1}{d}\sum_{q=1}^{d} e\left(\frac{mq}{d}\right) = -1,
\end{aligned}$$

we have

$$\left|S(\alpha\lambda^k,\wp)\right|^2 \le (d-1)^2 N(\wp)$$

and the theorem follows. □

Notes

For Theorem 2.1, see Hua [1].

Chapter 3

Weyl's Sum

3.1 Introduction

Let

$$f(\lambda) = \alpha_k \lambda^k + \cdots + \alpha_1 \lambda$$

be a polynomial of k-th degree with coefficients in J. The sum

$$S(f,T) = S\bigl(f(\lambda), \xi, T\bigr) = \sum_{\lambda \in P(T)} E\bigl(f(\lambda)\xi\bigr)$$

is called a Weyl's sum. Note that the range of the sum can be replaced by any finite set of integers. Before we state Weyl's inequality for $S(f,T)$ we begin with Siegel's generalisation of Dirichlet's theorem on rational approximations to real numbers.

Theorem 3.1 (Siegel). *Let $h > D^{1/n}$. Then, corresponding to any ξ, there exist an integer α and a number β of δ^{-1} such that*

$$\|\alpha\xi - \beta\| < \frac{1}{h}, \qquad 0 < \|\alpha\| \le h, \tag{3.1}$$

$$\max\bigl(h|\alpha^{(i)}\xi^{(i)} - \beta^{(i)}|, |\alpha^{(i)}|\bigr) \ge D^{-1/2}, \qquad 1 \le i \le n, \tag{3.2}$$

and

$$N\bigl((\alpha, \beta\delta)\bigr) \le D^{1/2}. \tag{3.3}$$

Theorem 3.2 (Weyl's inequality). *Let $G = 2^{k-1}$, and h be a number satisfying $T^{k-1} \ll h \ll T^k$. Then, for any $\epsilon > 0$,*

$$S(f,T) \ll T^{n+\epsilon} \left(\frac{1}{\|\alpha\|} + \frac{1}{T} + \frac{h}{T^k} \right)^{1/G},$$

where α denotes an integer satisfying the condition in Theorem 3.1 *for $\alpha_k\xi$, and the implicit constants depend on k, $\mathbf{K}$ and ϵ.*

We derive the following theorem from Theorem 3.2:

Theorem 3.3 (Schmidt). *Suppose that* $T \geq c_1(k, \mathbb{K}, \epsilon)$, $C \geq T^{n-1/G+\epsilon}$ *and* $|S(f,T)| \geq C$. *Then there exist a totally nonnegative integer* α *and an integer* β *such that*

$$\|\alpha\alpha_k\xi - \beta\| \ll \left(\frac{T^n}{C}\right)^G T^{-k+\epsilon}$$

and

$$0 < \|\alpha\| \ll \left(\frac{T^n}{C}\right)^G T^{\epsilon}.$$

3.2 Proof of Theorem 3.1

Let $x_1, \ldots, x_n$ and $y_1, \ldots y_n$ be real variables, and define

$$\alpha^{(i)} = \sum_{j=1}^{n} w_j^{(i)} x_j, \quad \beta^{(i)} = \sum_{j=1}^{n} \rho_j^{(i)} y_j, \quad \delta^{(i)} = \alpha^{(i)}\xi^{(i)} - \beta^{(i)}, \qquad 1 \leq i \leq n.$$

The expressions $\alpha^{(\ell)}$, $\delta^{(\ell)}$ together with the real and imaginary parts of $\sqrt{2}\alpha^{(m)}$, $\sqrt{2}\delta^{(m)}$ constitute $2n$ homogeneous linear forms in the $2n$ variables $x_1, \ldots, y_n$ with real coefficients and determinant

$$(-1)^n \left(\frac{-1}{i}\right)^{2r_2} \det\left(w_j^{(i)}\right) \det\left(\rho_j^{(i)}\right) = (-1)^{r_1+r_2}.$$

In view of Minkowski's theorem (see, for example, [Hua 1, §20.3]), there exist integral rational values of these variables, not all zero, so that $\|\alpha\| \leq h$ and $\|\delta\| < 1/h$ with α being an integer in $\mathbb{K}$ and β lies in the ideal δ^{-1}. If $\alpha = 0$, then $\|\beta\| < 1/h$. Hence $|N(\beta)| < h^{-n} \leq D^{-1}$, and consequently $\beta = 0$ which is impossible. Therefore $\alpha \neq 0$, and (3.1) is fulfilled.

Consider the finite set $\mathfrak{S}$ of all pairs α, β satisfying the conditions $1|\alpha$, $\delta^{-1}|\beta$ and (3.1). Choose α, β in $\mathfrak{S}$ so that the expression $\|\alpha\|$ attains its minimum b. Then $b \geq 1$. Put $(\alpha, \beta\delta)^{-1} = \mathfrak{q}$ and let κ denote a number of $\mathfrak{q}$. Since $(\alpha, \beta\delta)|\alpha$ and $\beta\delta(\alpha, \beta\delta)^{-1}$ are integrals, we have $1|\kappa\alpha$ and $\delta^{-1}|\kappa\beta$. The pair $\kappa\alpha = \hat{\alpha}$, $\kappa\beta = \hat{\beta}$ belong to $\mathfrak{S}$ whenever the conditions

$$\|\kappa\alpha\xi - \kappa\beta\| < \frac{1}{h}, \qquad 0 < \|\kappa\alpha\| \leq h, \tag{3.4}$$

are fulfilled, and then, in virtue of the definition of b,

$$\|\hat{\alpha}\| \geq b. \tag{3.5}$$

Let $\sigma_1, \ldots, \sigma_n$ be a basis of $\mathfrak{q}$ and $\gamma = \sum_{1 \leq j \leq n} x_j\sigma_j$, where x_j are real variables. The expressions $\gamma^{(\ell)}$ together with the real and imaginary parts of $\sqrt{2}\gamma^{(n)}$

constitutes n homogeneous linear forms in n variables with real coefficients and determinant $\pm N(\mathfrak{q})\sqrt{D}$. If $N(\mathfrak{q}) < D^{-1/2}$, then Minkowski's theorem implies the existence of a number $\kappa \in \mathfrak{q}$ such that $0 < \|\kappa\| < 1$. Hence (3.4) is satisfied by (3.1), and (3.5) leads to a contradiction because $\|\hat{\alpha}\| < \|\alpha\|$. It follows that $N(\mathfrak{q}^{-1}) \leq D^{1/2}$, and (3.3) is fulfilled.

It remains to prove also (3.2). This assertion is trivially true whenever $|\alpha^{(i)}| \geq D^{-1/2}$, $(1 \leq i \leq n)$. So we only have to investigate the case when

$$|\alpha^{(j)}| < D^{-1/2} \tag{3.6}$$

for some index $j \leq r_1 + r_2$, and we must prove that

$$h\left|\alpha^{(j)}\xi^{(j)} - \beta^{(j)}\right| \geq D^{-1/2}.$$

Since $\mathfrak{q}^{-1}$ is integral, we have $N(\mathfrak{q}) \leq 1$. Using Minkowski's theorem again we determine a number $\kappa \in \mathfrak{q}$ such that

$$0 < |\kappa^{(j)}| \leq D^{1/2}, \quad |\kappa^{(i)}| < 1, \qquad 1 \leq i \leq n, \quad i \neq j, \tag{3.7}$$

in the case $j \leq r_1$, and

$$0 < |\kappa^{(j)}| \leq D^{1/4}, \quad |\kappa^{(i)}| < 1, \qquad 1 \leq i \leq n, \quad i \neq j, j + r_2, \tag{3.8}$$

in the case $j > r_1$. In both cases we have $|\hat{\alpha}^{(j)}| < 1$ and $|\hat{\alpha}^{(j)}| < |\alpha^{(j)}| < 1$, $(i \neq j$ or $i \neq j, j + r_2)$ by (3.6). Since $\|\alpha\| = b \geq 1$, we obtain

$$\|\hat{\alpha}\| < \|\alpha\| = b$$

which contradicts (3.5) if the pair $\hat{\alpha}$, $\hat{\beta}$ were in $\mathfrak{S}$. Hence the conditions in (3.4) are not all satisfied. On the other hand, by (3.1), (3.6), (3.7) and (3.8),

$$0 < \|\kappa\alpha\| \leq h, \quad |\kappa^{(i)}|\,|\alpha^{(i)}\xi^{(i)} - \beta^{(i)}| < \frac{1}{h}, \qquad i \neq j \ \text{ or } \ i \neq j, j + r_2,$$

and so we have

$$h\left|\alpha^{(j)}\xi^{(j)} - \beta^{(j)}\right| \geq |\kappa^{(j)}|^{-1} \geq D^{-1/2}.$$

The lemma is proved. □

3.3 A Lemma on Units

Lemma 3.1. *Set $r = r_1 + r_2$. Let $\gamma_1, \ldots, \gamma_r$ be a set of real numbers satisfying*

$$\sum_{\ell} \gamma_\ell + 2\sum_{m} \gamma_m = 0. \tag{3.9}$$

Then there exists a totally nonnegative unit η of $\mathbb{K}$ such that

$$c_2^{-1}e^{\gamma_i} \le |\eta^{(i)}| \le c_2 e^{\gamma_i}, \qquad 1 \le i \le r,$$

where $c_2 = c_2(\mathbb{K})$.

Proof. We know by Dirichlet's unit theorem that there always exists a set of fundamental units $\sigma_1, \dots, \sigma_{r-1}$ for the group of totally nonnegative units of $\mathbb{K}$ (see, for example, [Hecke 1, §35]. Thus

$$\det(\log|\sigma_j^{(i)}|) \ne 0, \qquad 2 \le i \le r, \quad 1 \le j \le r-1. \tag{3.10}$$

Set

$$\xi^{(i)} = \sigma_1^{(i)x_1} \cdots \sigma_{r-1}^{(i)x_{r-1}}, \qquad 1 \le i \le r.$$

Further, let $c_2 = e^{c_3}$ where

$$c_3 = \frac{1}{2} \max_{1 \le i \le r} \sum_{1 \le j < r} \big|\log|\sigma_j^{(i)}|\big|.$$

Then

$$\prod_{\ell} \xi(\ell) \prod_{m} |\xi^{(m)}|^2 = \prod_{i=1}^{r-1} \left(\prod_{\ell} \sigma_i^{(\ell)} \prod_{m} |\sigma_i^{(m)}|^2 \right)^{x_i} = 1. \tag{3.11}$$

Consider the system of linear equations

$$x_1 \log|\sigma_1^{(i)}| + \cdots + x_{r-1} \log|\sigma_{r-1}^{(i)}| = \gamma_i, \qquad 1 \le i \le r. \tag{3.12}$$

It follows from (3.10) that (3.12), except for the equation corresponding to $i = 1$, has a unique solution, and this solution also satisfies the equation for $i = 1$ in (3.12) by (3.9) and (3.11).

Let $a_i\,(1 \le i < r)$ be rational integers such that

$$|a_i - x_i| \le \frac{1}{2}, \qquad 1 \le i < r.$$

We define a unit η of $\mathbb{K}$ by

$$\eta = \sigma_1^{a_1} \cdots \sigma_{r-1}^{a_{r-1}},$$

which gives

$$\big|\log|\eta^{(i)}| - \log|\xi^{(i)}|\big| = \big|\log|\eta^{(i)}| - \gamma_i\big| \le \sum_{1 \le j < r} |x_j - a_j|\,\big|\log|\sigma_j^{(i)}|\big| \le c_3.$$

The lemma follows. □

3.4 Asymptotic Formula for $N(\mathfrak{a}, T)$

Lemma 3.2. *Let $\mathfrak{a}$ be an integral ideal. Let $T_i\,(1 \le i \le n)$ be positive numbers such that $T_{m+r_2} = T_m$. Let $N(\mathfrak{a}, T)$ denote the number of elements $\nu \in \mathfrak{a}$ satisfying*

$$0 \le \nu^{(\ell)} \le T_\ell, \qquad |\nu^{(m)}| \le T_m. \tag{3.13}$$

Then

$$N(\mathfrak{a}, T) = \frac{(2\pi)^{r_2}}{\sqrt{D}\,N(\mathfrak{a})} T_1 \cdots T_n + O\left(\frac{T_0^{n-1}}{N(\mathfrak{a})^{1-1/n}}\right),$$

where $T_0 = \max\left(N(\mathfrak{a})^{1/n}, (T_1 \cdots T_n)^{1/n}\right)$.

Proof. 1) We shall prove that we can take a basis $\alpha_1, \ldots, \alpha_n$ of $\mathfrak{a}$ such that

$$|\alpha_j^{(i)}| \le c_4(K) N(\mathfrak{a})^{1/n}, \qquad 1 \le i, j \le n.$$

Let A be an ideal class containing $\mathfrak{a}$. Then $\mathfrak{a}$ is a product of a fixed ideal $\mathfrak{a}_0 \in A$, and a number α of $\mathbb{K}$: $\mathfrak{a} = \alpha\mathfrak{a}_0$. Let $\lambda_1, \ldots, \lambda_n$ be a basis of $\mathfrak{a}_0$. Then $\alpha_i = \alpha\lambda_i\,(1 \le i \le n)$ is a basis of $\mathfrak{a}$. Set

$$\gamma_i = \log|N(\alpha)|^{1/n} + \log|\alpha^{(i)}|^{-1}, \qquad 1 \le i \le n.$$

Then (3.9) holds, and it follows from Lemma 3.1 that there exists a totally nonnegative unit η such that

$$|N(\alpha)|^{1/n}|\alpha^{(i)}|^{-1} \ll |\eta^{(i)}| \ll |N(\alpha)|^{1/n}|\alpha^{(i)}|^{-1}, \qquad 1 \le i \le n.$$

Put $\alpha' = \alpha\eta$. Then

$$|N(\alpha')|^{1/n} \ll |\alpha'^{(i)}| \ll |N(\alpha')|^{1/n}, \qquad 1 \le i \le n.$$

We may use α' instead of α. Hence we may suppose that

$$|N(\alpha)|^{1/n} \ll |\alpha^{(i)}| \ll |N(\alpha)|^{1/n}, \qquad 1 \le i \le n.$$

Therefore

$$|\alpha_j^{(i)}| = |\alpha^{(i)}\lambda_j^{(i)}| \ll |N(\alpha)|^{1/n} \ll N(\mathfrak{a})^{1/n}.$$

2) For any $\mathbf{x} = (x_1, \ldots, x_n)$, where x_ℓ are real numbers and $x_m = \bar{x}_{m+r_2}$, we define $X_\ell(\mathbf{x}) = x_\ell$, $X_m(\mathbf{x}) = \mathrm{Re}(x_m)$, $X_{m+r_2}(\mathbf{x}) = \mathrm{Im}(x_m)$, and $\mathbf{X}(\mathbf{x}) = \left(X_1(\mathbf{x}), \ldots, X_n(\mathbf{x})\right)$. We set also

$$\mathbf{X}(\xi) = \left(\xi^{(1)}, \ldots, \xi^{(r_1)}, \mathrm{Re}(\xi^{(r_1+1)}), \ldots, \mathrm{Im}(\xi^{(r_1+r_2)})\right).$$

By Lemma 3.1, we can choose a totally nonnegative unit σ such that

$$(T_1 \cdots T_n)^{1/n} \ll |\sigma^{(i)}|T_i \ll (T_1 \cdots T_n)^{1/n}, \qquad 1 \le i \le n.$$

If we use $\sigma\nu$ instead of ν in $N(\mathfrak{a}, T)$, we see that

$$N(\mathfrak{a}, T) = N(\mathfrak{a}, |\sigma|T).$$

Therefore we may suppose that

$$(T_1 \cdots T_n)^{1/n} \ll T_i \ll (T_1 \cdots T_n)^{1/n}, \qquad 1 \le i \le n.$$

3) The vectors $\mathbf{X}(\alpha_1), \ldots, \mathbf{X}(\alpha_n)$ in E_n are linearly independent and they span a parallelopipe (v) whose volume v is $N(\mathfrak{a})\sqrt{D}2^{-r_2}$ and the diameter is less than $c_5(K)N(\mathfrak{a})^{1/n}$. In fact, let

$$\sigma_i = s_1\alpha_1^{(i)} + \cdots + s_n\alpha_n^{(i)}, \qquad 1 \le i \le n,\ 0 \le s_j \le 1,\ 1 \le j \le n,$$

and $\sigma_m = \eta_m + i\zeta_m$. The Jacobian of σ_ℓ, η_m, ζ_m with respect to s_j is

$$2^{-r_2}\left|\det(\alpha_j^{(i)})\right| = 2^{-2r_2}N(\mathfrak{a})\sqrt{D}.$$

Therefore

$$v = \int_{(v)} d\sigma\, d\eta\, d\zeta = 2^{-r_2}N(\mathfrak{a})\sqrt{D}\int_{U_n} ds = 2^{-r_2}N(\mathfrak{a})\sqrt{D},$$

where $d\sigma = \prod_\ell d\sigma_\ell$, $d\eta = \prod_m d\eta_m$, $d\zeta = \prod_m d\zeta_m$ and $ds = \prod_{1\le i\le n} ds_i$.

We define a domain (V_0) in E_n by the following conditions

$$(V_0) \qquad 0 \le x_\ell \le T_\ell, \quad x_m^2 + x_{m+r_2}^2 \le T_m^2, \quad -\pi \le \arg(x_m + ix_{m+r2}) \le \pi,$$

with $\mathbf{x} = (x_1, \ldots, x_n)$ being a point of E_n. Then $N(\mathfrak{a}, T)$ is equal to the number of lattice points in (V_0) with respect to the vectors $\mathbf{X}(\alpha_1), \ldots, \mathbf{X}(\alpha_n)$.

We denote by $\rho(\boldsymbol{\gamma}_1, \boldsymbol{\gamma}_2)$ the distance between the two points $\boldsymbol{\gamma}_1$ and $\boldsymbol{\gamma}_2$ in E_n and define two domains (V_1) and (V_2) in E_n as follows:

$$(V_1) = \left\{\boldsymbol{\gamma}_1 : \rho(\boldsymbol{\gamma}_1, \boldsymbol{\gamma}_2) \le c_5(K)N(\mathfrak{a})^{1/n} \quad \text{for any } \gamma_2 \in (V_0)\right\}$$

$$(V_2) = \left\{\boldsymbol{\gamma}_1 : \boldsymbol{\gamma}_1 \in (V_0),\ \rho(\boldsymbol{\gamma}_1, \boldsymbol{\gamma}_2) \le c_5(K)N(\mathfrak{a})^{1/n} \quad \text{for all } \gamma_2 \notin (V_0)\right\}$$

Denote by V_i the volumes of $(V_i)\,(i = 0, 1, 2)$. We have

$$V_1 - V_2 \ll T_0^{n-1}(T_0 + N(\mathfrak{a})^{1/n} - T_0 + N(\mathfrak{a})^{1/n}) \ll T_0^{n-1}N(\mathfrak{a})^{1/n}$$

and

$$V_0 = \int_{(V_0)} dx = \prod_\ell T_\ell \prod_m \int_{u^2+v^2\le T_m^2} du\, dv = \pi^{r_2}\prod_{i=1}^{n} T_i,$$

where $dx = \prod_{1\le i\le n} dx_i$. On the other hand

$$\frac{2^{r_2}}{N(\mathfrak{a})\sqrt{D}} V_2 \le N(\mathfrak{a}, T) \le \frac{2^{r_2}}{N(\mathfrak{a})\sqrt{D}} V_1.$$

The lemma follows. □

A similar result also holds if (3.15) is replaced by

$$|\nu^{(i)}| \le T_i, \qquad 1 \le i \le n. \tag{3.14}$$

We have the following:

Lemma 3.3. *Let $N_1(\mathfrak{a},T^n)$ denote the number of ν satisfying*

$$\begin{aligned} \nu \equiv \mu \ (\mathrm{mod}\ \mathfrak{a}), \quad 0 < \nu^{(\ell)} < c_6 N(\nu)^{1/n}, \\ |\nu^{(m)}| < c_6 N(\nu)^{1/n}, \quad N(\nu) \le T^n, \end{aligned} \tag{3.15}$$

where $c_6 = c_6(K)$ and μ is a given number in a residue class mod $\mathfrak{a}$. Then

$$N_1(\mathfrak{a},T^n) = \frac{c_7(K)}{N(\mathfrak{a})} T^n + O\left(\frac{T_0^{n-1}}{N(\mathfrak{a})^{1-1/n}}\right),$$

where $T_0 = \max\left(N(\mathfrak{a})^{1/n}, T\right)$.

In fact, it is clear that in Lemma 3.2 the condition $\nu \in \mathfrak{a}$ can be changed to $\nu \equiv \mu$ (mod $\mathfrak{a}$). The only difference between the proofs of these two lemmas is that we should use here the domain

$$\begin{aligned} 0 < x_\ell < c_6 X^{1/n}, \qquad x_m^2 + x_{m+r_2}^2 < c_6 X^{2/n}, \\ -\pi \le \arg(x_m + i x_{m+r_2}) \le \pi, \qquad X \le T^n, \end{aligned} \tag{v_0}$$

where $X = \prod_\ell x_\ell \prod_m (x_m^2 + x_{m+r_2}^2)$, instead of (V_0).

Lemma 3.4. *Suppose that $T^n \ge N(\mathfrak{a})$. We have*

$$\sum_\nu \frac{1}{N(\nu)} = \frac{c_7 n \log T}{N(\mathfrak{a})} + c_8(\mathfrak{a},\mu) + O(N(\mathfrak{a})^{-1+1/n} T^{-1}),$$

where ν runs over all integers satisfying (3.15).

Proof. Set

$$N_1(\mathfrak{a},T^n) = \frac{c_7 T^n}{N(\mathfrak{a})} + R(\mathfrak{a},T^n,\mu), \qquad R(\mathfrak{a},T^n,\mu) = O\left(\frac{T^{n-1}}{N(\mathfrak{a})^{1-1/n}}\right).$$

Then

$$\begin{aligned} \sum_\nu \frac{1}{N(\nu)} &= \sum_{t \le T^n} \frac{N_1(\mathfrak{a},t) - N_1(\mathfrak{a},t-1)}{t} \\ &= \frac{c_7}{N(\mathfrak{a})} \sum_{t \le T^n} \frac{1}{t} + \frac{R(\mathfrak{a},[T^n],\mu)}{[T^n]+1} + \left(\sum_{t=1}^{\infty} - \sum_{t > T^n}\right) \frac{R(\mathfrak{a},t,\mu)}{t(t+1)}. \end{aligned}$$

Since the series

$$\sum_{t=1}^{\infty} \frac{R(\alpha, t, \mu)}{t(t+1)}$$

is absolutely convergent, and

$$\sum_{t \leq T^n} \frac{1}{t} = n \log T + c + O\left(\frac{1}{T^n}\right),$$

the lemma follows. □

3.5 A Sum

Lemma 3.5. *Let $s \geq 1$ and $T_i\,(0 \leq i \leq s)$ be rational integers with $T_0 \geq 1$. Let M be the set of s-tuples $\mathbf{t} = (t_1, \ldots, t_s) \in \mathbb{Z}^s$ such that $\mathbf{t} \neq \mathbf{0}$ and*

$$T_i \leq t_i \leq T_i + T_0, \qquad 1 \leq i \leq s.$$

Let M_0 be a subset of M and define

$$S = \sum_{\mathbf{t} \in M_0} \min\left(\frac{1}{|t_1|}, \ldots, \frac{1}{|t_s|}\right).$$

Then

$$S \ll A^{1-1/s} \log(1 + T_0),$$

where A is the number of elements of M_0, with $A = 1$ when M_0 is empty.

Proof. We may assume without loss of generality that $T_i = 0$ for $1 \leq i \leq s$. Let $M_v\,(1 \leq v \leq s)$ be the subset of M_0 consisting of those $\mathbf{t}$ such that $t_v \geq t_i$ for $1 \leq i \leq s$. Write

$$A_v = \sum_{\mathbf{t} \in M_v} 1$$

and

$$S_v = \sum_{\mathbf{t} \in M_v} \min\left(\frac{1}{|t_1|}, \ldots, \frac{1}{|t_s|}\right).$$

We first prove that

$$S_1 \ll A_1^{1-1/s} \log(1 + T_0). \tag{3.16}$$

Denote by $\mathbf{u} = (u_1, \ldots, u_s)$ a point of E_s. For $t > 0$, we consider the domain $D(t)$ of E_s given by the conditions:

$$D(t): \quad t \geq u_1 > 0,\ u_1 \geq u_i \geq 0,\ 2 \leq i \leq s.$$

Let $M(t)$ be the set of rational integer points in $D(t)$ whose number we denote by $n(t)$. Moreover, let $M_0(t)$ be a subset of M_1 consisting of those $\mathbf{t}$ with $t \geq t_1$ and let $n_0(t)$ count the elements of $M_0(t)$. It is obvious that, for any t,

$$n(t) \geq n_0(t).$$

If $A_1 \ll 1$, then (3.16) is obvious. We assume therefore that $A_1 \gg 1$. Let t_0 be an integer such that

$$n(t_0) \leq A_1 = n_0(T_0) < n(t_0 + 1).$$

Then we can construct a mapping φ from $M_0(T_0)$ to $M_0(t_0+1)$ which satisfies the condition that, for every $\mathbf{t} \in M_0(T_0)$, the first coordinate of $\mathbf{t}$ is not less than that of the φ-image $\varphi(\mathbf{t}) = \mathbf{u}$ of $\mathbf{t}$, that is $u_1 \leq t_1$. Hence we have

$$\begin{aligned} S_1 = \sum_{\mathbf{t} \in M_1} \frac{1}{t_1} &= \sum_{\mathbf{t} \in M_0(T_0)} \frac{1}{t_1} \leq \sum_{\mathbf{u} \in M(t_0+1)} \frac{1}{u_1} \\ &\leq (t_0+2)^{s-1}\big(\log(t_0+2)+1\big). \end{aligned}$$

Since

$$A_1 \geq n(t_0) = 2^{s-1} + 3^{s-1} + \cdots + (t_0+1)^{s-1} \geq c_8(s)(t_0+1)^s,$$

we have

$$S_1 \ll A_1^{1-1/s} \log(t_0+2) \ll A_1^{1-1/s} \log(T_0+2).$$

In the same way we obtain

$$S_v \ll A_v^{1-1/s} \log(T_0+2).$$

Since $S \leq S_1 + \cdots + S_s$, the lemma follows. □

3.6 Mitsui's Lemma

The proof of Theorem 3.2 is based on the following important

Lemma 3.6 (Mitsui). *Let A, B, h be positive numbers satisfying $A \geq 1$, $h \gg 1$ and $1 \leq B \ll h$. Then, for any ξ,*

$$\begin{aligned} S = \sum_{\mu \in M(B)} \min_{1 \leq i \leq n} &\left(A, |1 - E(\xi \mu w_i)|^{-1}\right) \\ &\ll AB^n \left(\frac{1}{\|\alpha\|} + \frac{1}{B} + \frac{h \log h}{AB} + \frac{\log h}{A} \right), \end{aligned}$$

where α denotes an integer satisfying the conditions in Theorem 3.1 *for the ξ.*

This lemma will be proved in sections §3.8 and §3.9.

Lemma 3.7. *Let $x > 1$ and μ be a nonzero integer. Then the number of solutions of*

$$\xi_1 \cdots \xi_s = \mu, \qquad \|\xi_i\| \le x, \quad 1 \le i \le s,$$

is $O\big(|N(\mu)|^{\epsilon}(\log x)^{(r-1)(s-1)}\big)$.

Proof. Since $N(\xi_1)\cdots N(\xi_s) = N(\mu)$, it is known that the number of solutions of $\big(N(\xi_1),\ldots,N(\xi_s)\big)$ is $O\big(|N(\mu)|^{\epsilon}\big)$. For a given rational integer $c \ne 0$, the number of solutions of $N(\xi) = c$ with $\|\xi\| \le x$ is $O\big((\log x)^{r-1}\big)$. In fact, by Lemma 3.1, we can take an integer ξ_0 satisfying $N(\xi_0) = c$ and $|c|^{1/n} \ll \|\xi\| \ll |c|^{1/n}$. Let $\sigma_1,\ldots,\sigma_{r-1}$ be a set of fundamental units of K. Then any solution of $N(\xi) = c$ is of the form $\zeta^{u_0}\sigma_1^{u_1}\cdots\sigma_{r-1}^{u_{r-1}}\xi_0$ with rational integers u_i and a root of unity ζ. From

$$\left|\sigma_1^{(i)u_1}\cdots\sigma_{r-1}^{(i)u_{r-1}}\xi_0^{(i)}\right| \ll x, \qquad 1 \le i \le r-1,$$

we have

$$\sum_{j=1}^{r-1} u_j \log \sigma_j^{(i)} \ll \log x, \qquad 1 \le i \le r-1,$$

and so $u_j \ll \log x$, $(1 \le j \le r-1)$. The lemma is proved. □

Lemma 3.8. *Let μ be an integer. Then*

$$\sum_{\lambda \in P(T)} E(\xi\mu\lambda) \ll T^{n-1} \min_i \big(T, |1 - E(\xi\mu w_i)|^{-1}\big).$$

Proof. For a given integer w, we obtain by Lemma 3.2

$$\begin{aligned} E(\xi\mu w) \sum_{\lambda \in P(T)} E(\xi\mu\lambda) &= \sum_{\lambda \in P(T)} E\big(\xi\mu(w+\lambda)\big) \\ &= \sum_{\lambda \in P(T)} E(\xi\mu\lambda) + O(T^{n-1}). \end{aligned}$$

Therefore

$$\big(1 - E(\xi\mu\, w)\big) \sum_{\lambda \in P(T)} E(\xi\mu\lambda) \ll T^{n-1},$$

and

$$\sum_{\lambda \in P(T)} E(\xi\mu\lambda) \ll T^{n-1} \min\big(T, |1 - E(\xi\mu w)|^{-1}\big).$$

The lemma follows by setting $w = w_i$ for $1 \le i \le n$. □

Lemma 3.9. *Suppose that* $1 \leq q \leq k-1$. *We have*

$$|S(f,T)|^{2^q} \ll T^{(2^q-q-1)n} \sum_{\lambda_1,\ldots,\lambda_{q-1}} \cdots \sum_{\in M(2T)} \sum_{\lambda,\lambda_q} \sum_{\in P(T)} E(\lambda_1 \cdots \lambda_q h(\lambda, \lambda_1, \ldots, \lambda_q)\xi),$$

where

$$h(\lambda, \lambda_1, \ldots, \lambda_q) = k(k-1)\cdots(k-q+1)\alpha_k \lambda^{k-q} + \cdots$$

is a polynomial of degree $k-q$ *in* $\lambda, \lambda_1, \ldots, \lambda_q$ *with integer coefficients in* J.

Proof. By Hölder's inequality, we have

$$\begin{aligned} |S(f,T)|^{2^q} &= \left| \sum_{\lambda \in P(T)} \sum_{\lambda+\lambda_1 \in P(T)} E(f(\lambda+\lambda_1)\xi - f(\lambda)\xi) \right|^{2^{q-1}} \\ &\leq \left(\sum_{\lambda_1 \in M(2T)} \left| \sum_{\lambda \in M(T)} E(\lambda_1 g(\lambda, \lambda_1)\xi) \right| \right)^{2^{q-1}} \\ &\ll T^{(2^{q-1}-1)n} \sum_{\lambda_1 \in M(2T)} \left| \sum_{\lambda \in M(T)} E(\lambda_1 g(\lambda, \lambda_1)\xi) \right|^{2^{q-1}}, \end{aligned}$$

where $g(\lambda, \lambda_1) = k\alpha_k \lambda^{k-1} + \cdots$ is a polynomial of degree $k-1$ in λ, λ_1. By successive applications of Hölder's inequality we obtain

$$\begin{aligned} |S(f,T)|^{2^q} &\ll T^{(2^{q-2}-1)n} \sum_{\lambda_1 \in M(2T)} T^{(2^{q-1}-1)n} \times \\ &\quad \times \left| \sum_{\lambda \in P(T)} E(\lambda_1 \lambda_2 k(k-1)\alpha_k \lambda^{k-2}\xi + \cdots) \right|^{2^{q-2}} \\ &\ll \cdots \\ &\ll T^{(2^{q-1}+2^{q-2}+\cdots+2-q+1)n} \sum \cdots \sum E(\lambda_1 \cdots \lambda_q h(\lambda, \lambda_1, \ldots, \lambda_q)\xi), \end{aligned}$$

where, in the last line here, the conditions of summation are $\lambda_1, \ldots, \lambda_{q-1} \in M(2T)$ and $\lambda, \lambda + \lambda_q \in P(T)$. The lemma is proved. □

Proof of Theorem 3.2. By Lemma 3.9 we have

$$|S(f,T)|^G \ll T^{(G-k)n} \sum_{\lambda_1, \cdots,} \cdots \sum_{\lambda_{k-1} \in M(2T)} \left| \sum_{\lambda} E(\xi \mu \alpha_k \lambda) \right|,$$

where

$$\mu = k!\,\lambda_1 \dots \lambda_{k-1}. \tag{3.16}$$

Let $A(\mu)$ denote the number of solutions of (3.6). Then, by Lemmas 3.2 and 3.7, we have

$$A(\mu) = \begin{cases} O\left(T^{(k-2)n}\right), & \text{if } \mu = 0, \\ O(T^{\epsilon}), & \text{if } \mu \neq 0. \end{cases}$$

Hence

$$|S(f,T)|^G \ll T^{(G-2)n} + T^{(G-k)n+\epsilon} \sum_{\mu} \left| \sum_{\lambda} E(\xi \mu \alpha_k \lambda) \right|,$$

where the summation is extended over all integers μ, λ with $\mu \in M(k!\,GT^{k-1})$ and $\lambda \in P(T)$. Let $A = T$ and $B = k!\,(2T)^{k-1}$. Then, by Lemmas 3.6 and 3.8, we have

$$\begin{aligned} |S(f,T)|^G &\ll T^{(G-2)n} + T^{(G-k)n+\epsilon} \sum_{\mu} T^{n-1} \min_i \left(T,\ |1 - E(\xi\mu\alpha_k w_i)|^{-1}\right) \\ &\ll T^{(G-2)n} + T^{(G-k)n+\epsilon+n-1+1+(k-1)n} \left(\frac{1}{\|\alpha\|} + \frac{1}{T^{k-1}} + \frac{h \log h}{T^k} + \frac{\log h}{T} \right) \\ &\ll T^{Gn+2\epsilon} \left(\frac{1}{\|\alpha\|} + \frac{1}{T} + \frac{h}{T^k} \right). \end{aligned}$$

The theorem is proved. □

3.7 Proof of Theorem 3.3

Lemma 3.10. *For any nonzero integer σ, there exists a nonzero integer γ such that $\|\gamma\| \le c_8(K)$ and $\gamma\sigma \in P$.*

Proof. The lemma clearly holds when $r_1 = 0$. Suppose therefore that $r_1 > 0$. Let

$$c_8 = 4 \max_i \sum_{j=1}^{n} \left| w_j^{(i)} \right| \qquad \text{and} \qquad N_\ell = \frac{\sigma^{(\ell)}}{2|\sigma^{(\ell)}|} c_8.$$

Since the matrix $(w_j^{(\ell)})\,(1 \le \ell \le r_1,\ 1 \le j \le n)$ has rank r_1, we may suppose that $\det(w_j^{(i)}) \neq 0\,(1 \le \ell, j \le r_1)$. The system of linear equations

$$\sum_{j=1}^{r_1} w_j^{(\ell)} x_j = N_\ell, \qquad 1 \le \ell \le r_1$$

has a unique solution. Set $a_j = [x_j]$. Then we have an integer $r = \sum_{1 \le j \le r_1} a_j w_j$ satisfying $\gamma^{(\ell)}\sigma^{(\ell)} > 0$ and $\|\gamma\| \le c_8$. The lemma is proved. □

Proof of Theorem 3.3. We have

$$T^{k-\epsilon}\left(\frac{C}{T^n}\right)^G \geq T^{k-\epsilon}\left(\frac{T^{n-1/G+\epsilon}}{T^n}\right)^G \geq T^{k-1+\epsilon}$$

and

$$T^{k-\epsilon}\left(\frac{C}{T^n}\right)^G \ll T^{k-\epsilon}$$

by Lemma 3.2. Let

$$h = T^{k-\epsilon}\left(\frac{C}{T^n}\right)^G.$$

Then h satisfies the condition in Theorem 3.2. By Theorem 3.1, there exist an integer α' and a number β' of δ^{-1} satisfying

$$\|\alpha'\alpha_k\xi - \beta'\| < h^{-1} \qquad \text{and} \qquad 0 < \|\alpha'\| \leq h.$$

Since

$$T^{n+\epsilon/2G}\left(\frac{h}{T^k}\right)^{1/G} = T^{n+\epsilon/2G-\epsilon/G}\frac{C}{T^n} = CT^{-\epsilon/2G}$$

and

$$T^{n-1/G+\epsilon/2G} \leq CT^{\epsilon/2G-\epsilon} < CT^{-\epsilon/2G},$$

we have, by Theorem 3.2,

$$C \leq |S(f,T)| \leq T^{n+\epsilon/2G}\|\alpha'\|^{-1/G};$$

that is

$$0 < \|\alpha'\| \ll \left(\frac{T^n}{C}\right)^G T^{\epsilon/2}.$$

There exists a nonzero integer γ' such that $\|\gamma'\| \leq c_9(K)$ and $\gamma'\beta'$ is integral for every $\beta' \in \delta^{-1}$ (see, for example, [Hecke 1, §31]). By Lemma 3.10 there is a nonzero integer γ with $\|\gamma\| \leq c_8$ such that $\gamma\gamma'\alpha' \in P$. Let $\alpha = \gamma\gamma'\alpha'$ and $\beta = \gamma\gamma'\beta'$. The lemma follows. □

3.8 Proof of Lemma 3.6

We write

$$S(\xi\mu w_i) = e_i + d_i, \qquad 1 \leq i \leq n,$$

with rational integers e_i and $-\frac{1}{2} < d_i \leq \frac{1}{2}$, $1 \leq i \leq n$. Put

$$\theta = \sum_{i=1}^n e_i\rho_i \qquad \text{and} \qquad \zeta = \sum_{i=1}^n d_i\rho_i.$$

Then θ and ζ are functions of μ and we have

$$\delta^{-1} | \theta, \quad \mu\xi = \theta + \zeta, \quad |1 - E(\xi\mu w_i)| = 2|\sin \pi d_i|,$$

and

$$|X_i(\zeta)| \leq \sum_{j=1}^{n} |d_j \rho_j^{(i)}| \leq c_{10}(K) \sum_{j=1}^{n} |d_j|, \qquad 1 \leq i \leq n.$$

Hence

$$S \ll \sum_{\mu \in M(B)} \min_i (A, |d_i|^{-1}) \ll \sum_{\mu \in M(B)} \min_i (A, |X_i(\zeta)|^{-1}) = S^\star,$$

say. Put

$$c_{11} = n c_{10} D^{1/n}.$$

Then

$$|X_i(\zeta)| \leq \frac{1}{2} c_{11} D^{-1/n}.$$

Taking c_{10} suitably, we may assume that

$$c_{11} > D^{1/2}.$$

To each μ in the sum $S^\star$ we assign a vector

$$\mathbf{y}(\mu) = (B_1 X_1(\zeta), \ldots, B_n X_n(\zeta))$$

in E_n with

$$B_i = 2D^{1/n} |\alpha^{(i)}|, \qquad 1 \leq i \leq n.$$

All $\mathbf{y}(\mu)$ are contained in the parallelotope

$$\{\mathbf{x} : |x_i| \leq c_{11} |\alpha^{(i)}|, \ 1 \leq i \leq n\}.$$

Now we divide the set $(1, 2, \ldots, n)$ into three parts J_1, J_2, J_3 by the following conditions:

$$i \in J_1 \quad \text{if and only if} \quad \frac{B}{h} D^{1/n} \geq 2c_{11} |\alpha^{(i)}|,$$

$$j \in J_2 \quad \text{if and only if} \quad \frac{1}{2} \geq 2c_{11} |\alpha^{(j)}| > \frac{B}{h} D^{1/n},$$

$$q \in J_3 \quad \text{if and only if} \quad 2c_{11} |\alpha^{(q)}| > \frac{1}{2}.$$

The parts J_1 or J_2 may be empty, but J_3 is not empty on account of $\alpha \neq 0$ and $1 | \alpha$. Moreover, we have

$$|\alpha^{(j)}| \leq c_{11}^{-1} < D^{-1/2}, \qquad j \in J_1 + J_2.$$

Therefore, putting

$$\delta^{(i)} = \alpha^{(i)} \xi^{(i)} - \beta^{(i)}, \qquad 1 \leq i \leq n,$$

we have, by (3.2), that

$$(3.17) \qquad |\delta^{(j)}|^{-1} \le D^{1/2}h, \qquad j \in J_1 + J_2.$$

Set

$$(3.18) \qquad \tau_i = \frac{2B}{h}D^{1/n}\ (i \in J_1) \quad \text{and} \quad \tau_j = 4c_{11}|\alpha^{(j)}|\ (j \in J_2).$$

Since

$$\prod_{j\in J_1+J_2} \tau_j \prod_{q\in J_3} (c_{11}|\alpha^{(q)}|) \ge c_{11}^n|N(\alpha)| \ge c_{11}^n > 1,$$

we can choose positive numbers τ_q for $q \in J_3$ such that

$$c_{11}|\alpha^{(q)}| > \tau_q \ge 2^{-2n}, \quad \tau_{q+r_2} = \tau_q\,(q \ge r_1+1,\ q \in J_3) \quad \text{and} \quad \tau_1 \cdots \tau_n = 2^{-2n}.$$

Let $\mathbf{g} = (g_1, \ldots, g_n)$ be a point of $\mathbb{Z}^n$ and $B(\mathbf{g})$ a parallelotope in E_n which is defined as follows:

$$B(\mathbf{g}) = \{\mathbf{x} : \tau_i(g_i - 1/2) < x_i \le \tau_i(g_i + 1/2),\ 1 \le i \le n\}.$$

We shall consider $B(\mathbf{g})$ which contains at least one $\mathbf{y}(\mu)$. If $\mathbf{y}(\mu)$ and $\mathbf{y}(\mu_1)$ are contained in the same $B(\mathbf{g})$, then, by decomposing $\mu\xi$ and $\mu_1\xi$ as

$$\mu\xi = \theta + \zeta \qquad \text{and} \qquad \mu_1\xi = \theta_1 + \zeta_1,$$

we have

$$\left|B_i(X_i(\zeta) - X_i(\zeta_1))\right| < \tau_i, \qquad 1 \le i \le n,$$

so that

$$(3.19) \qquad \left|\alpha^{(i)}(X_i(\zeta) - X_i(\zeta_1))\right| < 2^{-1}D^{-1/n}\tau_i, \qquad 1 \le i \le n.$$

On the other hand, in view of (3.1) and (3.18), we have

$$(3.20) \qquad \left|\delta^{(i)}(X_i(\mu) - X_i(\mu_1))\right| < 2h^{-1}B \le D^{-1/n}\tau_i, \qquad 1 \le i \le n.$$

We now put

$$\kappa = \alpha(\theta - \theta_1) - \beta(\mu - \mu_1).$$

Then $\delta^{-1}|\kappa$ and

$$\kappa = \alpha(\mu - \mu_1)\xi - \alpha(\zeta - \zeta_1) - \beta(\mu - \mu_1) = -\alpha(\zeta - \zeta_1) + \delta(\mu - \mu_1).$$

Therefore, by (3.19) and (3.20), we have

$$\left|\kappa^{(\ell)}\right| < D^{-1/n}\tau_\ell + 2^{-1}D^{-1/n}\tau_\ell < 2D^{-1/n}\tau_\ell,$$
$$\left|\kappa^{(m)}\right| < 2D^{-1/n}\tau_m + D^{-1/n}\tau_m < 4D^{-1/n}\tau_m,$$

so that

$$|N(\kappa)| < 2^{2n}\tau_1 \cdots \tau_n D^{-1} = D^{-1}.$$

Since $\kappa \in \delta^{-1}$, this inequality implies $\kappa = 0$. Hence

$$\alpha(\zeta - \zeta_1) = \delta(\mu - \mu_1)$$

and

$$\beta(\mu - \mu_1) = \alpha(\theta - \theta_1) \in \alpha\delta^{-1}. \tag{3.21}$$

Let $\delta\beta\alpha^{-1} = \mathfrak{b}/\mathfrak{a}$, $(\mathfrak{a}, \mathfrak{b}) = 1$. Then (3.3) implies that

$$|N(\alpha)| = N(\mathfrak{a})N((\alpha, \beta\delta)) \leq D^{1/2}N(\mathfrak{a}).$$

It follows from (3.21) that

$$\delta\beta(\mu - \mu_1)\alpha^{-1} = \frac{(\mu - \mu_1)\mathfrak{b}}{\mathfrak{a}} = \mathfrak{L}$$

is an integral ideal. Since $(\mathfrak{a}, \mathfrak{b}) = 1$, we have $\mu - \mu_1 \in \mathfrak{a}$ and

$$\mathfrak{a}_1(\mu - \mu_1) \subset (\alpha),$$

where $\mathfrak{a}_1 = (\alpha)/\mathfrak{a}$. Therefore we have

$$\rho(\mu - \mu_1) \in (\alpha)$$

with a suitable element ρ of $\mathfrak{a}_1$ such that

$$\|\rho\| \leq c_{12}(K). \tag{3.22}$$

We denote by $W(\mathbf{g})$ the numbers μ such that

$$\mu \in M(B) \qquad \text{and} \qquad \mathbf{y}(\mu) \in B(\mathbf{g}). \tag{3.23}$$

If we choose a μ_1 satisfying (3.23), then we see from (3.22) that $W(\mathbf{g})$ does not exceed the number of integers ν such that

$$|\nu^{(i)}| \leq \max_{\mu} \left| \frac{\rho^{(i)}(\mu^{(i)} - \mu_1^{(i)})}{\alpha^{(i)}} \right|, \qquad 1 \leq i \leq n,$$

where μ runs through all the integers satisfying (3.23). If $j \in J_2 + J_3$, then

$$\max_{\mu} \left| \frac{\rho^{(j)}(\mu^{(j)} - \mu_1^{(j)})}{\alpha^{(j)}} \right| \ll \frac{B}{|\alpha^{(j)}|}.$$

If $i \in J_1$, then we have, by (3.17),

$$\max_{\mu} \left| \frac{\rho^{(i)}(\mu^{(i)} - \mu_1^{(i)})}{\alpha^{(i)}} \right| = \max_{\mu} \left| \frac{\rho^{(j)}(\zeta^{(j)} - \zeta_1^{(j)})}{\delta^{(j)}} \right| \ll h.$$

It follows by Lemma 3.2 that

$$W(\mathbf{g}) \ll 1 + h^{i_1} B^{n-i_1} \prod_{j \in J_2 + J_3} |\alpha^{(j)}|^{-1},$$

where i_1 is the number of elements of J_1. Set

$$W_0 = h^{i_1} B^{n-i_1} \prod_{j \in J_2 + J_3} |\alpha^{(j)}|^{-1}.$$

We can now write

$$S^\star = \sum_{\mathbf{g}} \sum_{\substack{\mathbf{y}(\mu) \in B(\mathbf{g}) \\ \mu \in M(B)}} \min_i \left(A, \frac{1}{|X_i(\zeta)|} \right),$$

where $\mathbf{g}$ runs over all points of $\mathbb{Z}^n$ for which each $B(\mathbf{g})$ contains at least one $\mathbf{y}(\mu)$. Let G_i be the least rational integer satisfying

$$c_{11}|\alpha^{(i)}| < \tau_i \left(G_i + \frac{1}{2} \right), \qquad 1 \le i \le n.$$

Since the i-th coordinate $B_i X_i(\zeta)$ of $\mathbf{y}(\mu)$ satisfies

$$\left| B_i X_i(\zeta) \right| \le c_{11} |\alpha^{(i)}|, \qquad 1 \le i \le n,$$

the range of $\mathbf{g}$ in $S^\star$ is given by the condition

$$|g_i| \le G_i, \qquad 1 \le i \le n.$$

Then we have

$$G_j = 0 \ (j \in J_1 + J_2) \qquad \text{and} \qquad 1 \le G_q \le 2c_{11}|\alpha^{(q)}|/\tau_q \ (q \in J_3).$$

Therefore we can write

$$S^\star = \sum_{\{g_q\}} \sum_{\substack{\mathbf{y}(\mu) \in B(\mathbf{g}) \\ \mu \in M(B)}} \min_i \left(A, \frac{1}{|X_i(\zeta)|} \right), \tag{3.24}$$

where $\sum_{\{g_q\}}$ means that this sum is taken over all g_q with $q \in J_3$.

3.9 Continuation

To prove Lemma 3.6 it suffices to show that $S^\star$ satisfies the inequality for S in Lemma 3.6, since S is dominated by $S^\star$. We divide the sum (3.24) into two parts:

$$S^\star = S_1 + S_2, \tag{3.25}$$

where S_1 denotes a sum of μ satisfying $\mathbf{y}(\mu) \in B(\mathbf{0})$ and $\mu \in M(B)$, and where S_2 is the sum of the remaining terms.

1) Estimation of S_1. If $J_1 + J_2 = \phi$, then $|N(\alpha)| \gg \|\alpha\|$ and

$$S_1 \ll (1+W_0)A \ll A + \frac{AB^n}{|N(\alpha)|} \ll A + \frac{AB^n}{\|\alpha\|}. \tag{3.26}$$

Now we assume that $J_1 + J_2 \neq \phi$. Since

$$\frac{1}{|z|} \leq \min\left(\frac{1}{|\mathrm{Re}(z)|}, \frac{1}{|\mathrm{Im}(z)|}, \right) \leq \frac{\sqrt{2}}{|z|}$$

holds for any complex number $z \neq 0$, we have

$$S_1 \ll \sum_{\substack{\mathbf{y}(\mu)\in B(\mathbf{a}) \\ \mu\in M(B)}} \min_i \left(A, \frac{1}{|\zeta^{(i)}|}\right).$$

If $W(\mathbf{0}) < 2$, then it is obvious that $S_1 \ll A$. We now assume that $W(\mathbf{0}) \geq 2$. If we fix a number μ_1 such that $\mathbf{y}(\mu_1) \in B(\mathbf{0})$, then other μ with $\mathbf{y}(\mu) \in B(\mathbf{0})$ satisfy the following condition:

$$\alpha(\zeta - \zeta_1) = \delta(\mu - \mu_1), \qquad \mu - \mu_1 \in \mathfrak{a}.$$

Therefore

$$\begin{aligned} S_1 &\ll \sum_{\substack{\mathbf{y}(\mu)\in B(\mathbf{o}) \\ \mu\in M(B)}} \min_{j\in J_1+J_2} \left(A, \frac{1}{|\zeta^{(j)}|}\right) \\ &\ll \sum_{\substack{\mu-\mu_1\in\mathfrak{a} \\ \mu\in M(B)}} \min_{j\in J_1+J_2} \left(A, \left|\frac{\delta^{(j)}}{\alpha^{(j)}}(\mu^{(j)} - \mu_1^{(j)}) + \zeta_1^{(j)}\right|^{-1}\right). \end{aligned}$$

Put

$$\xi_0^{(j)} = \begin{cases} \frac{\alpha^{(j)}}{\delta^{(j)}}\zeta_1^{(j)}, & \text{if} \quad j \in J_1 + J_2, \\ 0, & \text{if} \quad j \in J_3. \end{cases}$$

Then we have

$$S_1 \ll \sum_{\substack{\mu\in\mathfrak{a} \\ \mu\in M(2B)}} \min_{j\in J_1+J_2} \left(A, \frac{h}{|\mu^{(j)} + \xi_0^{(j)}|}\right) \ll \sum_{\substack{\mu\in\mathfrak{a} \\ \mu\in M(2B)}} \min_{j\in J_1+J_2} \left(A, \frac{h}{|X_j(\mu + \xi_0)|}\right).$$

Let $\mathbf{t} = (t_1, \ldots, t_n)$ be a point of $\mathbb{Z}^n$ and let

$$B^\star(\mathbf{t}) = \left\{\mathbf{x} : \frac{N(\mathfrak{a})^{1/n}}{3}\left(t_i - \frac{1}{2}\right) < x_i \leq \frac{N(\mathfrak{a})^{1/n}}{3}\left(t_i + \frac{1}{2}\right),\ 1 \leq i \leq n\right\}.$$

For every $B^\star(\mathbf{t})$ the number of $\mu \in \mathfrak{a}$ such that $\mathbf{X}(\mu + \xi_0) \in B^\star(\mathbf{t})$ is at most one. Otherwise it follows from $\mathbf{X}(\mu + \xi_0) \in B^\star(\mathbf{t})$ and $\mathbf{X}(\mu_1 + \xi_0) \in B^\star(\mathbf{t})$ that $\mu - \mu_1 \in \mathfrak{a}$ and $N(\mu - \mu_1) < N(\mathfrak{a})$; that is $\mu = \mu_1$. Moreover, we have

$$\min_{j\in J_1+J_2} \frac{1}{|X_j(\mu + \xi_0)|} \ll \min_{j\in J_1+J_2} \frac{1}{|t_j|\, N(\mathfrak{a})^{1/n}}$$

for $\mu + \xi_0$ such that $\mathbf{X}(\mu + \xi_0) \in B^\star(\mathbf{t})$. Therefore

$$S_1 = \sum_{\mathbf{t}} \min_{j \in J_1 + J_2} \left(A, \frac{h}{|t_j| \, N(\mathfrak{a})^{1/n}} \right), \tag{3.27}$$

where $\mathbf{t}$ runs over all integral vectors for which there exists $\mu \in \mathfrak{a}$ such that

$$\mathbf{X}(\mu + \xi_0) \in B^\star(\mathbf{t}) \qquad \text{and} \qquad \mathbf{X}(\mu) \in M(2B).$$

The range of $\mathbf{t}$ is given as follows:
(3.28)

$$T_j \le t_j \le T_j' \quad (j \in J_1 + J_2) \qquad \text{and} \qquad t_q \ll T_0 = 1 + \frac{B}{N(\mathfrak{a})^{1/n}} \quad (q \in J_3)$$

with $T_j' - T_j \ll T_0$. We divide the sum (3.27) into two parts:

$$\sum_{\mathbf{t}} = \Sigma_1 + \Sigma_2,$$

where Σ_1 is the sum taken over all $\mathbf{t}$ with $t_j = 0$ for all $j \in J_1 + J_2$, and Σ_2 consists of the other terms. Since J_3 and $J_1 + J_2$ are not empty we have

$$\Sigma_1 \ll A T_0^{n-1} \ll A B^{n-1}.$$

As for Σ_2, noting the existence of an index $j \in J_1 + J_2$ for which $t_j \ne 0$, we have

$$\Sigma_2 \ll \frac{h}{N(\mathfrak{a})^{1/n}} \sum_{\mathbf{t}}{}' \min_{j \in J_1 + J_2} \frac{1}{|t_j|},$$

where Σ' means a sum taken over all possible $\mathbf{t}$ with the range (3.28). We have, by Lemma 3.5,

$$\Sigma_2 \ll h T_0^{n-1} \log(1 + T_0) \ll B^{n-1} h \log h,$$

and so the sum of (3.27) satisfies

$$S_1 \ll A B^n \left(\frac{1}{B} + \frac{h \log h}{AB} \right).$$

Consequently, by (3.26),

$$S_1 \ll A B^n \left(\frac{1}{\|\alpha\|} + \frac{1}{B} + \frac{h \log h}{AB} \right). \tag{3.29}$$

2) Estimation of S_2. By the definition of $\mathbf{y}(\mu)$ we have, for μ such that $\mathbf{y}(\mu) \in B(\mathbf{g})$,

$$\min_{1 \le i \le n} \frac{1}{|X_i(\zeta)|} \ll \min_{1 \le i \le n} \frac{B_i}{\tau_i |g_i|} \ll \min_{t \in J_3} \frac{|\alpha^{(t)}|}{\tau_t |g_t|}$$

which gives

$$S_2 \ll \sum_{\{g_q\}\neq\{0\}} W(\mathbf{g}) \min_{t\in J_3} \frac{|\alpha^{(t)}|}{\tau_t|g_t|}.$$

First we assume that $W_0 > 1$. Then

$$S_2 \ll W_0 \sum_{\{g_q\}\neq\{0\}} \min_{t\in J_3} \frac{|\alpha^{(t)}|}{\tau_t|g_t|}.$$

Since $|g_i| \le G_i$, we have

$$\sum_{\{g_q\}\neq\{0\}} \min_{t\in J_3} \frac{|\alpha^{(t)}|}{\tau_t|g_t|} \ll \sum_{q\in J_3} \sum_{(q)}{}' \sum_{g_q=1}^{G_q} \frac{|\alpha^{(q)}|}{\tau_q|g_q|},$$

where $\Sigma'_{(q)}$ means a sum taken over g_t $(t \in J_3,\ t \neq q)$. Therefore the above expression is

$$\begin{aligned} &\ll \sum_{q\in J_3} \frac{|\alpha^{(q)}|}{\tau_q} \prod_{\substack{t\in J_3\\ t\neq q}} G_t \log(1+G_q) \\ &\ll \prod_{t\in J_3} \frac{|\alpha^{(t)}|}{\tau_t} \sum_{q\in J_3} \log(1+G_q) \ll \prod_{t\in J_3} \frac{|\alpha^{(t)}|}{\tau_t} \log h, \end{aligned}$$

and consequently

$$\begin{aligned} S_2 &\ll W_0 \prod_{t\in J_3} \frac{|\alpha^{(t)}|}{\tau_t} \log h \\ &\ll h^{i_1} B^{n-i_1} \prod_{t\in J_3} \tau_t^{-1} \prod_{j\in J_2} |\tau^{(j)}|^{-1} \log h \\ &\ll h^{i_1} B^{n-i_1} \prod_{t\in J_1+J_2} \tau_t \prod_{j\in J_2} |\tau^{(j)}|^{-1} \log h. \end{aligned}$$

Since

$$\tau_i \ll Bh^{-1}\ (i \in J_1) \qquad \text{and} \qquad \tau_j \ll |\alpha^{(j)}|\ (j \in J_2),$$

we have

$$S_2 \ll B^n \log h. \tag{3.30}$$

Finally we assume that $W_0 \le 1$. Then there exists a constant $c_{13}(K)$ such that

$$W(\mathbf{g}) \le c_{13}$$

for all $W(\mathbf{g})$. Let G_0 be the set of $\{g_q,\ q \in J_3\}$ such that

$$W(\mathbf{g}) \neq 0, \qquad \{g_q\} \neq \{0\}.$$

Then noting that

$$|\alpha^{(j)}| \leq h \ (1 \leq j \leq n) \qquad \text{and} \qquad \tau_q \geq 2^{-2n} \ (q \in J_3),$$

we have

$$S_2 = h \sum_{\{g_q\} \in G_0} \min_{q \in J_3} \frac{1}{|g_q|}.$$

The value of $|g_q|$ in G_0 does not exceed $G_q \ll h$. Therefore, by Lemma 3.5,

$$S_2 \ll hN^{1-1/n} \log h,$$

where N is the number of elements of G_0. The number N can be estimated by Lemma 3.2:

$$N \leq \sum_{\{g_q\} \neq \{0\}} W(\mathbf{g}) \leq \sum_{\mu \in M(B)} 1 \ll B^n.$$

Then we obtain

$$S_2 \ll hB^{n-1} \log h. \tag{3.31}$$

Combining (3.30) and (3.31) we have

$$S_2 \ll AB^n \left(\frac{\log h}{A} + \frac{h \log h}{AB} \right), \tag{3.32}$$

and the lemma follows by (3.25), (3.29) and (3.32).

Notes

Theorem 3.1: See Siegel [3].

Weyl's inequality was first generalised to any algebraic number field by Siegel [3,4]. Theorem 3.2 is an improved form of Siegel's result which was proved by Mitsui [1,2] using his Lemma 3.6. Roughly speaking, if $\mathbf{x}$ belongs to a supplementary domain, then $S(\lambda^k, \xi, T) \ll T^{n-1/(2^{k-1}+n)+\epsilon}$ can be established by Siegel's result, and $S(\lambda^k, \xi, T) \ll T^{n-1/2^{k-1}+\epsilon}$ by Theorem 3.2.

Theorem 3.3 was first proved by Schmidt for the rational field $\mathbb{Q}$ (see, for example, Schmidt [1]) and generalised to any algebraic number field K by Wang [1].

Lemma 3.5: See Mitsui [1,2].

Chapter 4

Mean Value Theorems

4.1 Introduction

Let

$$f(\lambda) = \alpha_k \lambda^k + \cdots + \alpha_1 \lambda$$

be a polynomial of k-th degree with coefficients in J, where $\alpha_i \in M\big(O(T^{k-i})\big)$, $1 \leq i \leq k$. Let

$$S(f(\lambda), \xi, T) = S(f, T) = \sum_{\lambda \in M'(T)} E(f(\lambda)\xi),$$

where we use $M'(T)$ to denote subset of $M(T)$ which may be distinct in different occurances.

Theorem 4.1 (Hua's inequality). *If* $1 \leq q \leq k$, *then*

$$\int_{U_n} |S(f,T)|^{2^q} dx \ll T^{(2^q - q)n + \epsilon} \tag{4.1}$$

where the implicit constant depends on k, K *and* ϵ.

Theorem 4.2 (Linnik's inequality). *We have*

$$\int_{U_n} |S(f,T)|^{8^{k-1}} dx \ll T^{(8^{k-1} - k)n}. \tag{4.2}$$

4.2 Proof of Theorem 4.1

If $q = 1$, the left hand side of (4.1) is equal to the number of solutions of the equation

$$f(\mu) = f(\nu), \qquad \mu, \nu \in M'(T).$$

By Lemmas 2.2 and 3.2, we see that the number of integers $\nu \in M'(T)$ is $O(T^n)$, and, for any given ν, the number of μ such that $f(\mu) = f(\nu)$ does not exceed k. The theorem follows. Suppose now that $1 \le q < k$, and that the theorem holds for q. By Lemma 3.9 with $M'(T)$ instead of $P(T)$, we have

$$|S(f,T)|^{2^q} \le c_1(q,K)\Big(T^{(2^q-1)n} + T^{(2^q-q-1)n} \times \\ \times \sum_{\lambda_1,\ldots,\lambda_{q-1}} \cdots \sum_{\in M(2T)} \sum_{\lambda,\lambda+\lambda_q} \sum_{\in M'(T)}^{\star} E(\lambda_1 \cdots \lambda_q h(\lambda,\lambda_1,\ldots,\lambda_q)\xi)\Big),$$

where $h(\lambda,\lambda_1,\ldots,\lambda_q)$ is a polynomial of degree $k-q$ in $\lambda,\lambda_1,\ldots,\lambda_q$ such that

$$h(\lambda,\lambda_1,\ldots,\lambda_q) = \beta_{k-q}\lambda^{k-q} + \cdots + \beta_1\lambda, \quad \beta_i \in M\big(O(T^{k-q-i})\big), \quad 1 \le i \le k-q,$$

and where $\star$ means that $\lambda_1 \cdots \lambda_q h(\lambda,\lambda_1,\ldots,\lambda_q) \ne 0$. Multiplying both sides by $|S(f,T)|^{2^q}$ and integrating with respect to $\mathbf{x}$ over U_n, we obtain

$$\int_{U_n} |S(f,T)|^{2^{q+1}} dx \le c_1(q,K)\Big(T^{(2^q-1)n} I_1 + T^{(2^q-q-1)n} I_2,\Big)$$

where

$$I_1 = \int_{U_n} |S(f,T)|^{2^q} dx$$

and I_2 is the integral over U_n of the sum over

$$E\Big(\big(\lambda_1 \cdots \lambda_q h(\lambda,\lambda_1,\ldots,\lambda_q) + f(\mu_1) + \cdots + f(\mu_{2^q-1}) - f(\nu_1) - \cdots - f(\nu_{2^q-1})\big)\xi\Big)$$

with the following associated summation conditions:

$$\sum_{\lambda_1,\ldots,\lambda_{q-1}} \cdots \sum_{\in M(2T)} \sum_{\lambda,\lambda+\lambda_q} \sum_{\in M'(T)}^{\star} \sum_{\mu_1,\ldots,\nu_{2^q-1}} \cdots \sum_{\in M'(T)} .$$

The contribution from I_2 does not exceed the number of solutions of the equation

$$(4.3)\quad \lambda_1 \cdots \lambda_q h(\lambda,\lambda_1,\ldots,\lambda_q) = f(\nu_1) + \cdots + f(\nu_{2^q-1}) - f(\mu_1) - \cdots - f(\mu_{2^q-1})$$

under the condition cited above. For a given set of integers $\nu_1,\ldots,\nu_{2^q-1},\mu_1,\ldots,\mu_{2^q-1}$ such that the right hand side of (4.3) is not zero, the number of solutions of (4.3) in $\lambda,\lambda_1,\ldots,\lambda_q$ does not exceed $O(T^\epsilon)$ by Lemmas 2.2 and 3.7. Therefore by induction, we have

$$\begin{aligned}\int_{U_n} |S(f,T)|^{2^{q+1}} dx &\ll T^{(2^q-1)n} \int_{U_n} |S(f,T)|^{2^q} dx + T^{(2^q-q-1)n+2^q n+\epsilon} \\ &\ll T^{(2^q-1)n+(2^q-q)n+\epsilon} + T^{(2^{q+1}-q-1)n+\epsilon} \\ &\ll T^{(2^{q+1}-q-1)n+\epsilon}.\end{aligned}$$

The theorem is proved. □

4.3 Proof of Theorem 4.2

Let $R(x)$ and $R(x,z)\,(1 \le x \le z)$ denote the numbers of solutions of the bilinear equation

$$\xi_1\zeta_1 + \xi_2\zeta_2 = \xi_3\zeta_3 + \xi_4\zeta_4$$

satisfying the respective conditions

$$\xi_i, \zeta_i \in M(x), \quad 1 \le i \le 4, \qquad \text{and} \qquad 0 \ne \xi_i \in M(x),\ \zeta_i \in M(z), \quad 1 \le i \le 4.$$

To prove Theorem 4.2 we shall need:

Lemma 4.1. *We have*

$$R(x) = O(x^{6n}) \qquad \text{and} \qquad R(x,z) = O\big((xz)^{3n}\big).$$

We shall prove Lemma 4.1 in §4.7. Since the lemma holds for $n = 1$ (see, for example, [Hua 1 §19.7]) we assume in the following sections that $n > 1$.

Proof of Theorem 4.2. Suppose that $k = 2$. The integral on the left hand side of (4.2) does not exceed the number of solutions of the equation

$$f(\lambda_1) + \cdots + f(\lambda_4) = f(\mu_1) + \cdots + f(\mu_4), \qquad \lambda_i, \mu_i \in M(T), \quad 1 \le i \le 4.$$

Set

$$\xi_i = (-1)^{i+1}(\lambda_i - \mu_i), \qquad \zeta_i = \alpha_2(\lambda_i + \mu_i) + \alpha_1, \qquad 1 \le i \le 4.$$

Then the number concerned does not exceed the number $R(O(T))$ of solutions of

$$\xi_1\zeta_1 + \xi_3\zeta_3 = \xi_2\zeta_2 + \xi_4\zeta_4, \qquad \xi_i, \zeta_i \in M(O(T)), \qquad 1 \le i \le 4.$$

By Lemma 4.1 we obtain $R(O(T)) = O(T^{6n})$, and the theorem follows.

Suppose now that $k \ge 3$ and that the theorem holds for $k-1$ instead of k. Then

$$\begin{aligned}\Big|\sum_{\lambda\in M'(T)} E(f(\lambda)\xi)\Big|^2 &= \sum_{\lambda\in M'(T)} 1 + \sum_{\lambda\in M'(T)} E(-f(\lambda)\xi) \sum_{\substack{\lambda+\delta\in M'(T)\\ \delta\ne 0}} E(f(\lambda+\delta)\xi)\\ &= \sum_{\lambda\in M'(T)} 1 + \sum_{0\ne\delta\in M'(2T)} \sum_{\lambda\in M'(T)} E\big(\delta g(\lambda,\delta)\xi\big),\end{aligned}$$

where

$$g(\lambda,\delta) = \frac{f(\lambda+\delta) - f(\lambda)}{\delta}, \qquad \delta \ne 0.$$

By Hölder's inequality,

$$\begin{aligned}
&\Big|\sum_{\lambda\in M'(T)} E(f(\lambda)\xi)\Big|^{2\cdot 8^{k-2}} \\
&\quad\ll T^{8^{k-2}n} + T^{(8^{k-2}-1)n} \sum_{0\neq\delta\in M(2T)} \Big|\sum_{\lambda\in M'(T)} E(\delta g(\lambda,\delta)\xi)\Big|^{8^{k-2}} \\
&\quad\ll T^{8^{k-2}n} + T^{(8^{k-2}-1)n} \sum_{0\neq\delta\in M(2T)} \sum_{\tau\in M(O(T^{k-1}))} a(\tau,\delta)E(\delta\tau\xi),
\end{aligned}$$

where

$$a(\tau,\delta) = \int_{U_n} \Big|\sum_{\lambda\in M'(T)} E(\delta g(\lambda,\delta)\xi)\Big|^{8^{k-2}} E(-\tau\xi)\,dx.$$

By induction we have

$$|a(\tau,\delta)| \le \int_{U_n} \Big|\sum_{\lambda\in M'(T)} E(\delta g(\lambda,\delta)\xi)\Big|^{8^{k-2}} dx \ll T^{(8^{k-2}-k+1)n},$$

and thus

$$\begin{aligned}
&\int_{U_n}\Big|\sum_{\lambda\in M'(T)} E(f(\lambda)\xi)\Big|^{8^{k-1}} dx \\
&\quad\ll T^{4\cdot 8^{k-2}n} + T^{4(8^{k-2}-1)n} \int_{U_n}\Big|\sum_{0\neq\delta\in M'(2T)} \sum_{\tau\in M(O(T^{k-1}))} a(\tau,\delta)E(\delta\tau\xi)\Big|^4 dx \\
&\quad\ll T^{4\cdot 8^{k-2}n} + T^{4(8^{k-2}-1)n} \sum\nolimits^{\star} a(\tau_1,\delta_1)a(\tau_2,\delta_2)a(\tau_3,\delta_3)a(\tau_4\delta_4) \\
&\quad\ll T^{4\cdot 8^{k-2}n} + T^{4(8^{k-2}-1)n+4(8^{k-2}-k+1)n} \sum\nolimits^{\star} 1,
\end{aligned}$$

where $\Sigma^\star$ denotes a sum over all integers $\delta_1,\ldots,\tau_4$ with

$$\delta_1\tau_1+\delta_2\tau_2=\delta_3\tau_3+\delta_4\tau_4,\quad 0\neq\delta_i\in M(2T),\quad \tau_i\in M(O(T^{k-1})),\quad 1\le i\le 4.$$

By Lemma 4.1 we have

$$\sum\nolimits^{\star} 1 \ll T^{3kn},$$

and the theorem follows. □

4.4 A Lemma on the set D

Let $\mathbf{x}=(x_1,\ldots,x_n)$, where the x_ℓ are real and $x_m=\bar{x}_{m+r_2}$. In §4.4–§4.6 we generalise the definitions of norm and $M(x)$ by letting $N(\mathbf{x})=x_1\cdots x_n$ and $M(x)=\{\mathbf{x}:|x_i|\le x,\ 1\le i\le n\}$. We use the notations:

$$D = \{\mathbf{x} : x_\ell > 0,\ |x_m| > 0\}, \qquad D(x) = D \cap M(x),$$
$$H(x) = \{\mathbf{x} : \mathbf{x} \in D,\ N(\mathbf{x}) \le x^n\}, \qquad \bar{H}(x) = \{\mathbf{x} : \mathbf{x} \in D,\ N(\mathbf{x}) = x^n\},$$
$$B_\eta(t) = \left\{\mathbf{x} : 0 < \frac{x_\ell}{\eta^{(\ell)}} \le \big(tN(\mathbf{x})\big)^{1/n},\ 0 < \left|\frac{x_m}{\eta^{(m)}}\right| \le \big(tN(\mathbf{x})\big)^{1/n},\right\}$$
$$C_\eta(x,t) = H(x) \cap B_\eta(t), \qquad \text{and} \qquad \bar{C}_\eta(x,t) = \bar{H}(x) \cap B_\eta(t),$$

where $t \ge 1$, and η, η', η'' denote totally nonnegative units of K. Also $\xi \in M(x)$ (or $D(x)$) means that $(\xi^{(1)}, \dots, \xi^{(n)}) \in M(x)$ (or $D(x)$).

Lemma 4.2. *There exists $c_2 = c_2(K)$ such that, if $t > c_2$, then*

$$D \subset \bigcup_\eta B_\eta(t).$$

Proof. If $\mathbf{x} \in D$, then $a\mathbf{x} \in D\,(a > 0)$, and if $\mathbf{x} \in B_\eta(t)$, then $a\mathbf{x} \in B_\eta(t)\,(a > 0)$. So it suffices to prove that, if $t > c_2$,

$$\bar{H}(1) \subset \bigcup_\eta \bar{C}_\eta(1,t).$$

Let

$$x'_j = \log x_j, \qquad 1 \le j \le n.$$

Then the x'_ℓ are real, and $x'_m = \log|x_m| + i\phi_m$, $x'_{m+r_2} = \log|x_m| - i\phi_m$, where $-\pi < \phi_m \le \pi$. The sets $H(1)$ and $\bar{C}_\eta(1,t)$ become the hyperplane

$$H' = \{\mathbf{x}' : x'_1 + \cdots + x'_n = 0\}$$

and the set

$$C'_\eta(t) = \{\mathbf{x}' : x'_\ell - \log\eta^{(\ell)} \le \tfrac{1}{n}\log t,\ x'_m - i\phi_m - \log|\eta^{(m)}| \le \tfrac{1}{n}\log t,$$
$$x'_{m+r_2} + i\phi_m - \log|\eta^{(m)}| \le \tfrac{1}{n}\log t\} \cap H'$$

respectively. Therefore it suffices to show that, if $t > c_2$,

$$H' \subset \bigcup_\eta C'_\eta(t).$$

The set $C'_\eta(t)$ can be obtained from $C'_1(t)$ by a translation of the vector

$$V(\eta) = \Big(\log\eta^{(1)}, \dots, \log\eta^{(r_1)}, \log|\eta^{(r_1+1)}|, \dots, \log|\eta^{(n)}|\Big).$$

By Dirichlet's unit theorem we see that every totally nonnegative unit can be represented uniquely by a set of totally nonnegative fundamental units $\sigma_1, \dots, \sigma_{r-1}$ as

$$\eta = \sigma_1^{u_1} \cdots \sigma_{r-1}^{u_{r-1}}, \qquad u_i = u_i(\eta), \qquad 1 \le i \le r-1,$$

where $r = r_1 + r_2$ and u_i are rational integers. Thus the set $\{V(\eta)\}$ is an $r-1$ dimensional lattice with basis $V(\sigma_1), \ldots, V(\sigma_{r-1})$ and

$$V(\eta) = u_1 V(\sigma_1) + \cdots + u_{r-1} V(\sigma_{r-1}).$$

Let $R_\eta(x)$ be the set

$$H' \cap \Big\{ \mathbf{x}' : \sum_{\ell} (x'_\ell - \log \eta^{(\ell)})^2 + \sum_{m} (x'_m - i\phi_m - \log|\eta^{(m)}|)^2 + \sum_{m} \left(x'_{m+r_2} + i\phi_m - \log|\eta^{(m)}|\right)^2 < x^2 \Big\}.$$

We have

$$R_\eta \left(\frac{\log t}{n} \right) \subset C'_\eta(t).$$

Since $\{V(\eta)\}$ is a lattice, it follows that

$$H' \subset \bigcup_{\eta} R_\eta \left(\frac{\log t}{n} \right)$$

if t is sufficiently large, and the lemma is proved. □

4.5 A Lemma on the set $D(x)$

We write

$$C_\eta(x, c_2+1) = C_\eta(x), \qquad W_i(x,h) = \{\mathbf{x} : |x_i| \le h\} \cap D(x),$$

$$\bar{C}_\eta(x, c_2+1) = \bar{C}_\eta(x), \qquad W(x,h) = \bigcup_i W_i(x,h).$$

Lemma 4.3. *For any natural number T, we have*

$$D(x) \subset W\left(c_3^2 x,\, c_3^2 x e^{-2c_4 T}\right) \cup \bigcup_{h(\eta)<T} C_\eta\left(c_3 x e^{-c_4 h(\eta)}\right),$$

where $h(\eta) = \max|u_i(\eta)|$, $c_3 = c_3(K)$ and $c_4 = c_4(K)$.

To prove this lemma we shall need:

Lemma 4.4. *Let $c_5 = \sqrt{n}\log(c_2+1)$. Then, for $\mathbf{x}' \in C'_\eta(c_2+1)$,*

$$|x'_\ell - \log\eta^{(\ell)}| \le c_5, \quad |x'_m - i\phi_m - \log|\eta^{(m)}|| \le c_5, \quad |x'_{m+r_2} + i\phi_m - \log|\eta^{(m)}|| \le c_5.$$

Proof. We shall prove the following stronger assertion. For any $t > 1$,

$$C'_\eta(t) \subset R_\eta(\sqrt{n}\log t).$$

Since the figures of $C'_\eta(t)$, and also of $R_\eta(\sqrt{n}\log t)$, are congruent for each η, it suffices to show that

$$C'_1(t) \subset R_1(\sqrt{n}\log t).$$

If $\mathbf{x}' \in C'_1(t)$, then $x'_\ell \le \frac{\log t}{n}$, $x'_m - i\phi_m \le \frac{\log t}{n}$ and $x'_{m+r_2} + i\phi_m \le \frac{\log t}{n}$. It follows by $x'_1 + \cdots + x'_n = 0$ that $x'_\ell \ge -\frac{n-1}{n}\log t$, $x'_m - i\phi_m \ge -\frac{n-1}{n}\log t$ and $x'_{m+r_2} + i\phi \ge -\frac{n-1}{n}\log t$. Therefore

$$|x'_\ell| \le \log t, \qquad |x'_m - i\phi_m| \le \log t, \qquad |x'_{m+r_2} + i\phi_m| \le \log t,$$

and

$$\sum_\ell x'^2_\ell + \sum_m \left((x'_m - i\phi_m)^2 + (x'_{m+r_2} + i\phi_m)^2\right) \le n\log^2 t,$$

that is $\mathbf{x}' \in R_1(\sqrt{n}\log t)$. The lemma is proved. □

We define for each η, the polyhedron

$$P_\eta = \left\{\mathbf{x}' : \mathbf{x}' = \sum_{i=1}^{r-1}(u_i(\eta) + z_i)V(\sigma_i),\ -\tfrac{1}{2} \le z_j \le \tfrac{1}{2}\ (1 \le j \le r-1)\right\}$$

Let $c_6(K)$ be the distance from origin to the boundary of P_η and $c_7(K)$ the length of diagonal of P_η. Then the length of $V(\eta)$ satisfies

$$2c_6 h(\eta) \le |V(\eta)| \le c_7 h(\eta). \tag{4.4}$$

Lemma 4.5. *There exists c_4 such that, for each η,*

$$\max\log|\eta^{(i)}| \ge c_4 h(\eta) \qquad \text{and} \qquad \min\log|\eta^{(i)}| \le -c_4 h(\eta).$$

Proof. By (4.4) we have

$$n\max\log^2|\eta^{(i)}| \ge |V(\eta)|^2 \ge \left(2c_6 h(\eta)\right)^2.$$

We may suppose without loss of generality that $\big|\log|\eta^{(i)}|\big| = \big|\log|\eta^{(1)}|\big|$. Then

$$\big|\log|\eta^{(1)}|\big| \ge \frac{2c_6}{\sqrt{n}}h(\eta).$$

1) If $\log|\eta^{(1)}| \ge 0$, then $\max\log|\eta^{(i)}| = \log|\eta^{(1)}| \ge \frac{2c_6}{\sqrt{n}}h(\eta)$.
2) If $\log|\eta^{(1)}| < 0$, then $\log|\eta^{(1)}| \le -\frac{2c_6}{\sqrt{n}}h(\eta)$. Since $N(\eta) = 1$ and $n \ge 2$, we have

$$0 \le (n-1)\max\log|\eta^{(i)}| + \log|\eta^{(1)}| \le (n-1)\max\log|\eta^{(i)}| - \frac{2c_6}{\sqrt{n}}h(\eta).$$

3) If $\log|\eta^{(1)}| < 0$, then $\min\log|\eta^{(i)}| = \log|\eta^{(1)}| \le -\frac{2c_6}{\sqrt{n}}h(\eta)$.
4) If $\log|\eta^{(1)}| \ge 0$, then $\log|\eta^{(1)}| \ge \frac{2c_6}{\sqrt{n}}h(\eta)$. It follows from $N(\eta) = 1$ that

$$0 \ge (n-1)\min\log|\eta^{(i)}| + \log|\eta^{(1)}| \ge (n-1)\min\log|\eta^{(i)}| + \frac{2c_6}{\sqrt{n}}h(\eta).$$

Set $c_4 = 2c_6/\big(\sqrt{n}(n-1)\big)$. The lemma follows. □

Let $x_i = ye^{x'_i}$, $1 \le i \le n$, where $y > 0$. Then H' becomes $\bar{H}(y)$, and $C'_\eta(t)$ becomes $\bar{C}_\eta(y,t)$. By Lemma 4.4 we have

Lemma 4.6. *If* $\mathbf{x} \in \bar{C}_\eta(y)$, *then*

$$\eta^{(\ell)}c_3^{-1}y \le x_\ell \le \eta^{(\ell)}c_3y, \qquad |\eta^{(m)}|c_3^{-1}y \le |x_m| \le |\eta^{(m)}|c_3y,$$

where $c_3 = e^{c_5}$.

Lemma 4.7. *Suppose that* $x > 0$. *Then*

$$D(x) \subset \bigcup_\eta C_\eta(c_3xv_\eta^{-1}),$$

where $v_\eta = \max|\eta^{(i)}|$.

Proof. It follows from Lemma 4.2 that, for any $\mathbf{x} \in D(x)$, there is an $\eta = \eta(\mathbf{x})$ such that $\mathbf{x} \in B_\eta(c_2+1)$. It suffices to show that, for this η,

$$\mathbf{x} \in C_\eta(y), \qquad y = c_3xv_\eta^{-1}.$$

Set

$$b = b(\mathbf{x}, y) = y(x_1\cdots x_n)^{-1/n}.$$

Then $\mathbf{y} = b\mathbf{x} \in \bar{C}_\eta(y)$. Since

$$C_\eta(x,t) = \{\mathbf{x} : x = a\mathbf{y},\ \mathbf{y} \in \bar{C}_\eta(x,t),\ 0 < a \le 1\}, \tag{4.5}$$

it suffices to show that $b \ge 1$. By Lemma 4.6,

$$v_\eta c_3^{-1}y \le \max|y_i| = b\max|x_i| \le bx.$$

The assertion follows by the definition of y. The lemma is proved. □

Proof of Lemma 4.3. If $\mathbf{x} \in W(x,h)$, then $a\mathbf{x} \in W(x,h)\,(0 < a \le 1)$. Therefore, by (4.5), Lemmas 4.5 and 4.7, it suffices to show that, for $h(\eta) \ge T$,

$$\bar{C}_\eta(c_3cv_\eta^{-1}) \subset W\big(c_3^2x,\ c_3^2xe^{-2c_4T}\big).$$

Suppose that $\mathbf{x} \in \bar{C}_\eta(c_3xv_\eta^{-1})$ and $h(\eta) \ge T$. Then, by Lemma 4.6,

$$0 < x_\ell, \qquad |x_m| \le v_\eta c_3(c_3xv_\eta^{-1}) = c^2x;$$

that is, $\mathbf{x} \in D(c_3^2 x)$. On the other hand, it follows from Lemmas 4.5 and 4.6 that

$$\min |x_i| \le c_3^2 x v_\eta^{-1} \min |\eta^{(i)}| \le c_3^2 x e^{-2c_4 T}.$$

Therefore $\mathbf{x} \in W\left(c_3^2 x,\ c_3^2 x e^{-2Tc_4}\right)$, and the lemma follows. □

4.6 Fundamental Lemma

Let $r(\nu; x, z)$ and $r(x, z)$ denote the numbers of solutions in J of the bilinear equations

$$\xi_1\zeta_1 + \xi_2\zeta_2 = \nu, \qquad \xi_i \in M(x),\ \zeta_i \in M(z), \quad i = 1, 2,$$

and

$$\xi_1\zeta_1 + \xi_2\zeta_2 = 0, \qquad 0 \ne \xi_i \in M(x),\ \zeta_i \in M(z), \quad i = 1, 2,$$

respectively.

Lemma 4.8. *We have*

$$r(\nu; x, z) \ll (xz)^n \sum_{\mathfrak{a} | \nu} \frac{1}{N(\mathfrak{a})}, \qquad \textit{if} \qquad \nu \ne 0,$$

$$r(0; x, z) \ll (xz)^{3n/2} + (x + z)^{2n}$$

and

$$r(x, z) \ll (xz)^{3n/2} + x^{2n}.$$

We denote by $r_{\eta'\eta''}(\nu; x_1, x_2, z)$ the number of solutions in J of the equations

$$(4.6) \quad \xi_1\zeta_1 + \xi_2\zeta_2 = \nu, \quad \xi_1 \in C_{\eta'}(x_1),\ \xi_2 \in C_{\eta''}(x_2),\ \zeta_i \in M(z), \quad i = 1, 2.$$

Lemma 4.9. *If $\nu \ne 0$ and $1 < x_1 \le x_2 \le z$, then*

$$r_{\eta'\eta''}(\nu; x_1, x_2, z) \ll x_1^n z^n \left(1 + \log \frac{x_2}{x_1}\right) \sum_{\mathfrak{a} | \nu} \frac{1}{N(\mathfrak{a})}.$$

Proof. Suppose that $\xi_1, \xi_2, \zeta_1, \zeta_2$ is a solution of (4.6). Let $\mathfrak{a} = (\xi_1, \xi_2)$ be the ideal generated by ξ_1 and ξ_2. Then $\mathfrak{a} | \nu$ and

$$N(\mathfrak{a}) \ll x_1^n, \qquad N(\mathfrak{a}) \le N(\xi_i), \qquad i = 1, 2.$$

For other solutions $\xi_1, \xi_2, \zeta_1', \zeta_z'$ of (4.6), we have

$$(4.7) \qquad \xi_1(\zeta_1', -\zeta_1) = -\xi_2(\zeta_2' - \zeta_2).$$

Here ζ_2' is uniquely determined if ξ_1, ξ_2, ζ_1' are given. The same situation holds for ζ_1' and ξ_1, ξ_2, ζ_2'. By (4.7) we obtain

$$\zeta_1' \equiv \zeta_1 \ (\mathrm{mod}\ \frac{(\xi_2)}{\mathfrak{a}}) \quad \text{and} \quad \zeta_2' \equiv \zeta_2 \ (\mathrm{mod}\ \frac{(\xi_1)}{\mathfrak{a}})$$

For any integer $\gamma \in C_\eta(x)$ with $\gamma \equiv \mu \ (\mathrm{mod}\ \mathfrak{a})$, set $\gamma' = \gamma\eta^{-1}$. Then $\gamma' \in C_1(x)$ and $\gamma' \equiv \mu\eta^{-1} \ (\mathrm{mod}\ \mathfrak{a})$. This transformation is bijective, so that Lemmas 3.3 and 3.4 remain true when $C_1(x)$ is replaced by $C_\eta(x)$. Since $N(\xi_i) \ll x_i^n \leq z^n$, it follows that the numbers of ζ_1' and ζ_2' do not exceed

$$O\left(\frac{z^n N(\mathfrak{a})}{N(\xi_2)}\right) \qquad \text{and} \qquad O\left(\frac{z^n N(\mathfrak{a})}{N(\xi_1)}\right)$$

respectively. Therefore

$$\begin{aligned} r_{\eta'\eta''}(\nu; x_1, x_2, z) &\ll \sum_{\mathfrak{a}|\nu} \sum_{\substack{\xi_1 \in C_\eta'(x_1) \\ (\xi_1, \xi_2) = \mathfrak{a}}} \sum_{\xi_2 \in C_\eta''(x_2)} \frac{z^n N(\mathfrak{a})}{\max\left(N(\xi_1), N(\xi_2)\right)} \\ &\ll z^n \sum_{\mathfrak{a}|\nu} N(\mathfrak{a})(S_1 + S_2 + S_3), \end{aligned}$$

where

$$S_1 = \sum_{\substack{\xi_1 \in C_{\eta'}(x_1) \\ N(\xi_1) \leq N(\xi_2)}}^{\star} \sum_{\xi_2 \in C_{\eta''}(x_1)}^{\star} \frac{1}{\max\left(N(\xi_1), N(\xi_2)\right)},$$

$$S_2 = \sum_{\substack{\xi_1 \in C_{\eta'}(x_1) \\ N(\xi_1) > N(\xi_2)}}^{\star} \sum_{\xi_2 \in C_{\eta''}(x_1)}^{\star} \frac{1}{\max\left(N(\xi_1), N(\xi_2)\right)}$$

and

$$S_3 = \sum_{\xi_1 \in C_{\eta'}(x_1)}^{\star} \sum_{\xi_2 \in C_{\eta''}(x_2) - C_{\eta''}(x_1)}^{\star} \frac{1}{\max\left(N(\xi_1), N(\xi_2)\right)}$$

in which we use the notation $C'' - C' = \{\mathbf{x} : \mathbf{x} \in C'', \mathbf{x} \notin C'\}$, and $\Sigma^\star$ denotes a sum with $\mathfrak{a}|\xi_1$ or $\mathfrak{a}|\xi_2$.

By Lemma 3.3 we have

$$\begin{aligned} S_1 &\ll \sum_{\xi_2 \in C_{\eta''}(x_2)}^{\star} \frac{1}{N(\xi_2)} \sum_{\xi_1 \in C_{\eta'}\left(N(\xi_2)^{1/n}\right)}^{\star} 1 \\ &\ll \sum_{\xi_2 \in C_{\eta''}(x_1)}^{\star} \frac{1}{N(\xi_2)} \frac{N(\xi_2)}{N(\mathfrak{a})} \ll \frac{x_1^n}{N(\mathfrak{a})^2}. \end{aligned}$$

Similarly

$$S_2 \ll \frac{x_1^n}{N(\mathfrak{a})^2}.$$

Finally, by Lemmas 3.3 and 3.4, we obtain

$$S_3 \ll \sum_{\xi_2 \in C_{\eta''}(x_2) - C_{\eta'}(x_1)}^{\star} \frac{1}{N(\xi_2)} \sum_{\xi_1 \in C_{\eta''}(x_1)}^{\star} 1$$

$$\ll \sum_{\xi_2 \in C_{\eta''}(x_2) - C_{\eta''}(x_1)}^{\star} \frac{1}{N(\xi_2)} \frac{x_1^n}{N(\mathfrak{a})}$$

$$\ll \left(\sum_{\xi_2 \in C_{\eta''}(x_2)} - \sum_{\xi_2 \in C_{\eta''}(x_1)} \right) \frac{x_1^n}{N(\xi_2) N(\mathfrak{a})} \ll \frac{x_1^n}{N(\mathfrak{a})^2} \left(\log \frac{x_2}{x_1} + 1 \right).$$

The lemma is proved. □

Let $r_1(\nu; x, z)$ denote the number of solutions in J of the equation

$$\xi_1 \zeta_1 + \xi_2 \zeta_2 = \nu, \qquad \xi_i \in D(x), \quad \zeta_i \in M(z), \quad i = 1, 2.$$

Lemma 4.10. *If* $\nu \neq 0$ *and* $1 < x \leq z$, *then*

$$r_1(\nu; x, z) \ll x^n z^n \sum_{\mathfrak{a} | \nu} \frac{1}{N(\mathfrak{a})}.$$

Proof. If $T \geq T(x)$, where

$$T(x) = \frac{n}{2c_4} \log x + \frac{n \log c_3}{c_4},$$

then there is no integer in the set $W(c_3^2 x, \, c_3^2 x e^{-2c_4 T})$. Otherwise, if the set contains an integer ξ, then

$$0 < N(\xi) < (c_3^2 x)^{n-1} (c_3^2 x e^{-2c_4 T}) = (c_3^2 x)^n e^{-2c_4 T} \leq 1.$$

This leads to a contradiction. Hence, by Lemma 4.3 with c_4/n instead of c_4, we obtain

(4.8)
$$r_1(\nu; x, z) \leq \sum_{h(\eta') \leq T(x)} \sum_{h(\eta'') < T(x)} r_{\eta' \eta''}\left(\nu; \, c_3 x e^{-c_4 h(\eta')/n}, \, c_3 x e^{-c_4 h(\eta'')/n}, \, z\right).$$

If $h(\eta) < T(x)$, then

$$c_3 x e^{-c_4 h(\eta)/n} \geq \sqrt{x} > 1,$$

and therefore we can apply Lemma 4.9 to (4.8). We use the notations $S_\nu = \sum_{\mathfrak{a} | \nu} 1/N(\mathfrak{a})$, S_1 is the sum in the right hand side of (4.8) with η', η'' satisfying $h(\eta') \leq h(\eta'')$, and S_2 is the sum with the other terms. Then

$$r_1(\nu; x, z) \leq S_1 + S_2.$$

Since the number of units η satisfying $h(\eta) = H$ is at most $(r-1)(2H+1)^{r-2}$, we have, by Lemma 4.9, that

$$\begin{aligned}
S_1 &\ll x^n z^n S_\nu \sum_{h(\eta')\le h(\eta'')<T(x)} e^{-c_4 h(\eta'')/n}(1+h(\eta'')-h(\eta')) \\
&\ll x^n z^n S_\nu \sum_{m'\le m''<T(x)} (m'm'')^{r-2}(1+m''-m')e^{-c_4 m''/n} \\
&\ll x^n z^n S_\nu \sum_{m'=0}^{\infty}\sum_{m''=m'}^{\infty} (1+m'')^{2n}e^{-c_4 m''/n} \\
&\ll x^n z^n S_\nu \sum_{m'=0}^{\infty} e^{-c_4 m'/(2n)} \\
&\ll x^n z^n S_\nu .
\end{aligned}$$

Similarly

$$S_2 \ll x^n z^n S_\nu .$$

The lemma is proved. □

Let $\{\mathbf{t}\}$ be the set of 2^{r_1} vectors $(\tau_1, \dots, \tau_{r_1})$, where $\tau_\ell = \pm 1$. Write $\mathbf{x}^{(\mathbf{t})} = (\tau_1 x_1, \dots, \tau_{r_1} x_{r_1}, x_{r_1+1}, \dots, x_n)$. We use the notations:

$$\begin{aligned}
D^{(\mathbf{t})} &= \{\mathbf{x} : \mathbf{x}^{(\mathbf{t})} \in D\}, \\
D^{(\mathbf{t})}(x) &= \{\mathbf{x} : \mathbf{x}^{(\mathbf{t})} \in D(x)\}, \\
B_\eta^{(\mathbf{t})}(t) &= \{\mathbf{x} : \mathbf{x}^{(\mathbf{t})} \in B_\eta(t)\},
\end{aligned}$$

Let $r^{(\mathbf{t}_1,\mathbf{t}_2)}(\nu; x, z)$ be the number of solutions to the equation

$$\xi_1\zeta_1 + \xi_2\zeta_2 = \nu, \qquad \xi_i \in D^{(\mathbf{t}_i)}(x),\ \zeta_i \in M(z), \quad i = 1, 2.$$

Then $r^{(\mathbf{1},\mathbf{1})}(\nu; x, z) = r_1(\nu; x, z)$. We can prove similarly to the proof of Lemma 4.10, with $|N(\xi_i)|$ instead of $N(\xi_i)\,(i = 1, 2)$, that

$$r^{(\mathbf{t}_1,\mathbf{t}_2)}(\nu; x, z) \ll x^n z^n S_\nu, \qquad \nu \neq 0, \quad 1 < x \le z. \tag{4.9}$$

Let $r^{(\mathbf{0},\mathbf{t})}(\nu; x, z)$ denote the number of solutions of the equation

$$\xi_1\zeta_1 + \xi_2\zeta_2 = \nu, \quad \xi_1 = 0,\ \xi_2 \in D^{(\mathbf{t})}(x),\ \zeta_i \in M(z),\ i = 1, 2.$$

We may define similarly $r^{(\mathbf{t},\mathbf{0})}(\nu; x, z)$ and $r^{(\mathbf{0},\mathbf{0})}(\nu; x, z)$.

Lemma 4.11. *Let $x > 1$, $z > 1$ and $\nu \neq 0$. Then*

$$r^{(\mathbf{0},\mathbf{t})}(\nu; x, z) = r^{(\mathbf{t},\mathbf{0})}(\nu; x, z) \ll x^n z^n, \tag{4.10}$$

and

$$r^{(\mathbf{0},\mathbf{0})}(\nu; x, z) = 0. \tag{4.11}$$

Proof. By symmetry, in order to prove (4.10), it suffices to show that $r^{(\mathbf{0},\mathbf{t})}(\nu;x,z)=O\big((xz)^n\big)$. The possible values of ζ_1 are $O(z^n)$, and the number of solutions of $\xi_2\zeta_2=\nu$ is $O(x^n)$, since $\xi_2\in D^{(\mathbf{t})}(x)$. The assertion follows. The equation (4.11) is obvious, since $\nu\neq 0$ and $\xi_1=\xi_2=0$. The lemma is proved. □

Proof of Lemma 4.8. 1) Since

$$r(\nu;x,z)=\sum r^{(\mathbf{t}_1,\mathbf{t}_2)}(\nu;x,z),$$

where, for $i=1,2$, the vectors $\mathbf{t}_i$ run over $\{\mathbf{t}\}$ and $\mathbf{0}$, we have, by (4.9) and Lemma 4.10, that

$$r(\nu;x,z)\ll (xz)^n S_\nu,\qquad \nu\neq 0,\quad 1<x\leq z.$$

Note that $r(\nu;x,z)=r(\nu;z,x)$, so that the condition $1<x\leq z$ can be eleiminated.

2) Now we are going to prove the result for $r(0;x,z)$. By Lemma 3.2, the number of pairs ξ_2,ζ_2 with $\mu=-\xi_2\zeta_2\neq 0$ and $\xi_2\zeta_2=0$ are at most $O\big((xz)^n\big)$ and $O\big((x+z)^n\big)$. Given $\mu\neq 0$, it follows by Lemma 3.7 that the number of solutions of $\xi_1\zeta_1=\mu$ is $O\big((xz)^{n/2}\big)$. Since the number of solutions of $\xi_1\zeta_1=0$ is $O\big((x+z)^n\big)$, we have $r(0;x,z)\ll (xz)^{3n/2}+(x+z)^{2n}$.

3) Finally we prove that $r(x,z)\ll (xz)^{3n/2}+x^{2n}$. The number of pairs ξ_2,ζ_2 such that $\mu=-\xi_2\zeta_2\neq 0$ and $\xi_2\zeta_2=0$ are $O\big((xz)^n\big)$ and $O(x^n)$ respectively. If $\mu\neq 0$, then the number of solutions of $\mu=\xi_1\zeta_1$ is $O\big((xz)^{n/2}\big)$. Since the number of solutions of $\xi_1\zeta_1=0$ is $O(x^n)$, the assertion follows. □

4.7 Proof of Lemma 4.1

1) For given $\mu=\xi_1\zeta_1+\xi_2\zeta_2$, the number of solutions of

$$\xi_1\zeta_1+\xi_2\zeta_2=\xi_3\zeta_3+\xi_4\zeta_4,\qquad \xi_i,\zeta_i\in M(x),\ 1\leq i\leq 4$$

is $r(\mu;x,x)^2$. Since $\mu\in M(2x^2)$, we have, by Lemma 4.8,

$$\begin{aligned}R(x)&\leq\sum_{\mu\in M(2x^2)} r(\mu;x,x)^2\\&\ll x^{6n}+x^{4n}\sum_{0\neq\mu\in M(2x^2)}\sum_{\mathfrak{a}|\mu}\frac{1}{N(\mathfrak{a})}\sum_{\mathfrak{b}|\mu}\frac{1}{N(\mathfrak{b})}\\&\ll x^{6n}+x^{4n}\sum_{N(\mathfrak{a})\leq 2^n x^{2n}}\frac{1}{N(\mathfrak{a})}\sum_{N(\mathfrak{b})\leq 2^n x^{2n}}\frac{1}{N(\mathfrak{b})}\sum_{\substack{0\neq\mu\in M(2x^2)\\ [\mathfrak{a},\mathfrak{b}]|\mu}}1.\end{aligned}$$

By Lemma 3.2,

$$\sum_{\substack{0 \neq \mu \in M(2x^2) \\ [\mathfrak{a},\mathfrak{b}] \mid \mu}} 1 \ll \frac{x^{2n}}{N[\mathfrak{a},\mathfrak{b}]}.$$

Since

$$N[\mathfrak{a},\mathfrak{b}] \geq \left(N(\mathfrak{a})N(\mathfrak{b})\right)^{1/2},$$

we have

$$\sum_{\mathfrak{a}}\sum_{\mathfrak{b}} \frac{1}{N(\mathfrak{a})N(\mathfrak{b})N[\mathfrak{a},\mathfrak{b}]} \leq \left(\sum_{\mathfrak{a}} \frac{1}{N(\mathfrak{a})^{3/2}}\right)^2 \ll 1,$$

and therefore

$$R(x) \ll x^{6n}.$$

2) We have

$$\begin{aligned} R(x,z) &\leq r(x,z)^2 + \sum_{0 \neq \mu \in M(2xz)} r(\mu; x, z)^2 \\ &\ll (xz)^{3n} + (xz)^{2n} \sum_{0 \neq \mu \in M(2xz)} \sum_{\mathfrak{a} \mid \mu} \frac{1}{N(\mathfrak{a})} \sum_{\mathfrak{b} \mid \mu} \frac{1}{N(\mathfrak{b})} \\ &\ll (xz)^{3n}. \end{aligned}$$

The lemma is proved. □

Notes

Theorem 4.1 was first proved by Hua for $\mathbb{Q}$ (see, for example, [Hua 4]), and generalised to $\mathbb{K}$ by Ayoub [1] and Birch [2]. Hua's inequality was improved recently by Vaughan [1,2] and Heath-Brown [3], and it may be possible that a corresponding improvement can also be obtained for $\mathbb{K}$.

Theorem 4.2 is a variant of Hua's inequality which was first proved by Linnik [1] for $\mathbb{Q}$ (see also Khintchine [1] or [Hua 1]). This has been generalised to $\mathbb{K}$ by Rieger [1,2].

Chapter 5

The Circle Method in Algebraic Number Fields

5.1 Introduction

Let h and t be real numbers satisfying

$$h > 2Dt, \qquad t > 1.$$

For any $\gamma \in K$, we can determine uniquely two integral ideals $\mathfrak{a}, \mathfrak{b}$ such that

$$\gamma\delta = \frac{\mathfrak{b}}{\mathfrak{a}}, \qquad (\mathfrak{a}, \mathfrak{b}) = 1.$$

We write $\gamma \to \mathfrak{a}$ for this. Let $\Gamma(t)$ be the set consisting of those $\gamma = x_1\rho_1 + \cdots + x_n\rho_n$ satisfying

$$\mathbf{x} \in U_n \; : \; x_i \in \mathbb{Q}, \, 1 \leq i \leq n, \; \gamma \to \mathfrak{a} \text{ and } N(\mathfrak{a}) \leq t^n.$$

For every $\gamma \in \Gamma(t)$ subject to $\gamma \to \mathfrak{a}$, we define the basic domain

$$B_\gamma = \Big\{\mathbf{x} : \mathbf{x} \in U_n, \prod_{i=1}^{n} \max\left(h|\xi^{(i)} - \gamma_0^{(i)}|, t^{-1}\right) \leq N(\mathfrak{a})^{-1} \text{ for some } \gamma_0 \equiv \gamma \pmod{\delta^{-1}}\Big\}.$$

Note that B_γ is not a rectangle.

Lemma 5.1. *Let $\gamma_1, \gamma_2 \in \Gamma(t)$ with $\gamma_1 \neq \gamma_2$. Then*

$$B_{\gamma_1} \cap B_{\gamma_2} = \phi.$$

Proof. Suppose that there is a $\xi \in B_{\gamma_1} \cap B_{\gamma_2}$, that is

$$\prod_{i=1}^{n} \max\left(h|\xi^{(i)} - \gamma_{0j}^{(i)}|, t^{-1}\right) \leq N(\mathfrak{a}_j)^{-1},$$

where $\gamma_{0j} \to \mathfrak{a}_j$ and $\gamma_{0j} \equiv \gamma_j \pmod{\delta^{-1}}$, $j = 1, 2$. For simplicity, we set $\gamma_0 j = \gamma_j$, $j = 1, 2$ and write

$$\max\left(h|\xi^{(i)} - \gamma_j^{(i)}|, t^{-1}\right) = \sigma_j^{(i)}, \quad 1 \leq i \leq n, \; j = 1, 2.$$

Then

$$\prod_{i=1}^{n} \sigma_j^{(i)} \le N(\mathfrak{a}_j)^{-1}, \qquad \max_i \sigma_j^{(i)-1} \le t, \quad t = 1, 2$$

and

$$\begin{aligned} |\gamma_1^{(i)} - \gamma_2^{(i)}| &\le |\xi^{(i)} - \gamma_1^{(i)}| + |\xi^{(i)} - \gamma_2^{(i)}| \le h^{-1}(\sigma_1^{(i)} + \sigma_2^{(i)}) \\ &= h^{-1}\sigma_1^{(i)}\sigma_2^{(i)}(\sigma_1^{(i)-1} + \sigma_2^{(i)-1}) \le 2h^{-1}\sigma_1^{(i)}\sigma_2^{(i)} t. \end{aligned}$$

Therefore

$$N(\mathfrak{a}_1\mathfrak{a}_2)\,|N(\gamma_1 - \gamma_2)| \le (2h^{-1}t)^n < D^{-1}.$$

On the other hand, $\mathfrak{a}_1\mathfrak{a}_2(\gamma_1 - \gamma_2)\delta$ is an integral ideal, and thus

$$N(\mathfrak{a}_1\mathfrak{a}_2)\,|N(\gamma_1 - \gamma_2)| \ge N(\delta^{-1}) = D^{-1}.$$

This gives a contradiction, and therefore the lemma is proved. □

We set

$$B = \bigcup_{\gamma \in \Gamma(t)} B_\gamma,$$

and define the supplementary domain S of B with respect to U_n by

$$S = U_n - B.$$

We call this the *Farey division* of U_n into B and S with respect to (h, t).

Lemma 5.2. *Let α be an integer. Then*

$$\int_{U_n} E(\alpha\xi)\,dx = \begin{cases} 1, & \text{if } \alpha = 0, \\ 0, & \text{if } \alpha \ne 0. \end{cases}$$

Proof. The first assertion is obvious. Suppose then that $\alpha \ne 0$. Let $\alpha = y_1 w_1 + \cdots + y_n w_n$, where $y_i \in \mathbb{Z}$ are not all zero. Then

$$\int_{U_n} E(\alpha\xi)\,dx = \prod_{i=1}^{n}\prod_{j=1}^{n} \int_0^1 e(x_i y_j S(\rho_i w_j))\,dx_i = \prod_{i=1}^{n} \int_0^1 e(x_i y_i)\,dx_i = 0.$$

The lemma is proved. □

For nonzero integers $\alpha_i\,(1 \le i \le s)$ we define:

$$A = \max_i \|\alpha_i\|, \quad \xi - \gamma = \zeta,$$

$$G_i(\gamma) = N(\mathfrak{a}_i)^{-1} \sum_{\lambda(\mathfrak{a}_i)} E(\alpha_i \lambda^k \gamma),$$

$$T_i(\zeta, T) = \int_{P(T)} E(\alpha_i \eta^k \zeta)\,dy$$

$$S(\alpha_i\lambda^k, \xi, T) = S_i(\xi, T) = \sum_{\lambda \in P(T)} E(\alpha_i \lambda^k \xi), \qquad 1 \le i \le s,$$

where $\gamma \in \Gamma(t)$ and $\gamma\alpha_i \to \mathfrak{a}_i$, $1 \le i \le s$. We also write

$$G(\gamma) = \prod_{i=1}^{s} G_i(\gamma),$$

$$I(\zeta, T) = \prod_{i=1}^{s} I_i(\zeta, T), \qquad I(\zeta) = \prod_{i=1}^{s} I_i(\zeta, 1),$$

$$S(\xi, T) = \prod_{i=1}^{s} S_i(\xi, T), \quad J(\mu) = \int_{E_n} I(\zeta) E(-\mu\zeta)\, dx,$$

and

$$\mathfrak{S}(\nu) = \sum_{\gamma} G(\gamma) E(-\nu\gamma),$$

where $\mu = \nu T^{-k}$ and γ runs over a complete residue system mod δ^{-1}. The integral $J(\mu)$ is called a *singular integral* and the series $\mathfrak{S}(\nu)$ a *singular series*.

For given integers α_i and ν, consider the diophantine equation

$$\nu = \alpha_1 \lambda_1^k + \cdots + \alpha_s \lambda_s^k, \qquad \lambda_i \in P(T), \quad 1 \le i \le s. \tag{5.1}$$

This includes the two most interesting cases: 1) When $\alpha_1 = \cdots = \alpha_s = 1$, we have *Waring's problem.* 2) When $\nu = 0$, (5.1) is an additive equation. The study of diophantine inequalities for forms depends on the estimation for small solutions of additive equations. Let $R_s(\nu)$ be the number of solutions of (5.1). Then it follows from Lemma 5.2 that

$$R_s(\nu) = \int_{U_n} S(\xi, T) E(-\nu\xi)\, dx = \int_B S(\xi, T) E(-\nu\xi)\, dx + \int_S S(\xi, T) E(-\nu\xi)\, dx.$$

The integral over B gives the principal term of $R_s(\nu)$, and the main difficulty lies in the estimation of the integral over S which reduces to the estimation of Weyl's sums. This method was first introduced by Siegel, and it is now called *Siegel's generalised circle method.* When $\mathbb{K} = \mathbb{Q}$, we have the ordinary circle method of Hardy and Littlewood.

We shall write

$$a = \frac{1}{4} + \frac{1}{4kn}, \qquad t = T^{1-a} \quad \text{and} \quad h = T^{k-1+a}. \tag{5.2}$$

The object of this chapter is to prove:

Theorem 5.1. *If $s \ge 4kn$, then*

$$\int_B S(\xi, T) E(-\nu\xi)\, dx = \mathfrak{S}(\nu) J(\mu) T^{(s-k)n} + O(\Phi T^{(s-k)n}),$$

where

$$\Phi = \left(A^s + A^{2sn/k}\right) T^{-a} + A^{2sn^2/k} T^{-(1-a)n} + T^{-2^{-k}}.$$

Theorem 5.2. *If* $s \geq \max(4kn,\, 2^k+1)$, *then*

$$R_s(\nu) = \mathfrak{S}(\nu)J(\mu)T^{(s-k)n} + O(\Phi T^{(s-k)n}).$$

Remark. Intead of (5.1), we may consider the equation

$$\nu = f_1(\lambda_1) + \cdots + f_s(\lambda_s), \qquad \lambda_i \in P(T), \quad 1 \leq i \leq s,$$

where

$$f_i(\lambda) = \alpha_{ki}\lambda^k + \cdots + \alpha_{1i}\lambda, \qquad 1 \leq i \leq s,$$

are polynomials with coefficients in J, and $\alpha_{ki} \neq 0$, $1 \leq i \leq s$; see Subbarao and Wang [1].

5.2 Lemmas

Lemma 5.3. *Let* α *be a nonzero integer, and* $\tau = \alpha\zeta$. *Then*

$$\int_{P(T)} E(\alpha\eta^k\zeta)\,dy \ll \prod_{i=1}^{n} \min\left(T, |\tau^{(i)}|^{-1/k}\right).$$

Proof. Let

$$\eta^{(\ell)} = u_\ell, \qquad \eta^{(m)} = u_m e^{i\varphi_m}. \tag{5.3}$$

The Jacobian of y_i with respect to u_ℓ, u_m, φ_m is

$$2^{r_2}D^{-1/2}\prod_m u_m.$$

Set $\tau^{(m)} = |\tau^{(m)}|e^{i\psi_m}$. Then

$$|\tau^{(m)}|u_m^k\left(e^{i(k\varphi_m+\psi_m)} + e^{-i(k\varphi_m+\psi_m)}\right) = 2|\tau^{(m)}|u_m^k\cos(k\varphi_m+\psi_m).$$

Let $\cos(k\varphi_m+\psi_m) = \sin\theta_m$. We have

$$\int_{P(T)} E(\alpha\eta^k\zeta)\,dy = 2^{r_2}D^{-1/2}\times$$

$$\prod_\ell \int_0^T e(\tau^{(\ell)}u_\ell^k)\,du_\ell \prod_m \int_0^T \int_{-\pi}^{\pi} e(2|\tau^{(m)}|u_m^k\sin\theta_m)u_m\,du_m\,d\theta_m.$$

Let $|\tau^{(\ell)}|u^k = v$. Then

$$\int_0^T e\big(\tau^{(\ell)}u^k\big)\,du = |\tau^{(\ell)}|^{-1/k}k^{-1}\int_0^{|\tau^{(\ell)}|T^k} e(\pm v)v^{1/k-1}dv$$
$$\ll \min\big(T, |\tau^{(\ell)}|^{-1/k}\big).$$

If $k > 2$ and $\sin\theta \neq 0$, then

$$\int_0^T e(2|\tau^{(m)}|u^k\sin\theta)u\,du = \frac{1}{2}\int_o^{T^2} e(2|\tau^{(m)}|v^{k/2}\sin\theta)dv$$
$$= k^{-1}(2|\tau^{(m)}\sin\theta|)^{-2/k}\int_0^{2|\tau^{(m)}\sin\theta|T^k} e(\pm w)w^{2/k-1}dw$$
$$\ll |\tau^{(m)}\sin\theta|^{-2/k},$$

and therefore

$$\int_{-\pi}^{\pi}\int_0^T e(2|\tau^{(m)}u^k\sin\theta)u\,du\,d\theta \ll \min\left(T^2, |\tau^{(m)}|^{-2/k}\int_{-\pi}^{\pi}|\sin\theta|^{-2/k}d\theta\right)$$
$$\ll \min\big(T^2, |\tau^{(m)}|^{-2/k}\big).$$

Suppose that $k = 2$. We suppose that $|\tau^{(m)}|^{-1} \ll T^2$, since otherwise the lemma holds trivially. Then

$$\int_{-\pi}^{\pi}\int_0^T e(2|\tau^{(m)}|u^2\sin\theta)u\,du\,d\theta$$
$$= \frac{1}{2}\left(\int_0^{\pi}\frac{e(2|\tau^{(m)}|T^2\sin\theta)}{4\pi i|\tau^{(m)}|\sin\theta}d\theta + \int_{-\pi}^{0}\frac{e(2|\tau^{(m)}|T^2\sin\theta)}{4\pi i|\tau^{(m)}|\sin\theta}d\theta\right)$$
$$= \int_0^{\pi}\frac{\sin(4\pi|\tau^{(m)}|T^2\sin\theta)}{4\pi|\tau^{(m)}|\sin\theta}d\theta$$
$$= |2\pi\tau^{(m)}|^{-1}\int_0^{\pi/2}\frac{\sin(4\pi|\tau^{(m)}|T^2\sin\theta)}{\sin\theta}d\theta$$
$$= |2\pi\tau^{(m)}|^{-1}\int_0^{1}\frac{\sin(4\pi|\tau^{(m)}|T^2x)}{x\sqrt{1-x^2}}dx$$
$$= |2\pi\tau^{(m)}|^{-1}(I+J),$$

where

$$I = \int_0^{1/\sqrt{2}}\frac{\sin(4\pi|\tau^{(m)}|T^2x)}{x\sqrt{1-x^2}}dx \quad\text{and}\quad J = \int_{1/\sqrt{2}}^{1}\frac{\sin(4\pi|\tau^{(m)}|T^2x)}{x\sqrt{1-x^2}}dx,$$

and since

$$J \ll \int_{1/\sqrt{2}}^{1}\frac{dx}{\sqrt{1-x^2}} \ll 1,$$

it remains to show that $I \ll 1$.

Let L be the integer satisfying

$$\frac{L}{4|\tau^{(m)}|T^2} \leq \frac{1}{\sqrt{2}} < \frac{L+1}{4|\tau^{(m)}|T^2}.$$

Then $L \gg 1$ and

$$\begin{aligned} I &= \sum_{\ell=1}^{L} \int_{\ell/4|\tau^{(m)}|T^2}^{(\ell-1)/4|\tau^{(m)}|T^2} \frac{\sin(4\pi|\tau^{(m)}|T^2 x)}{x\sqrt{1-x^2}} dx + O(1) \\ &= \frac{1}{4|\tau^{(m)}|T^2} \sum_{\ell=1}^{L} (-1)^{\ell-1} \int_0^1 \frac{\sin \pi z}{\frac{\ell-1+z}{4|\tau^{(m)}|T^2} \sqrt{1 - \left(\frac{\ell-1+z}{4|\tau^{(m)}|T^2}\right)^2}} dz + O(1). \end{aligned}$$

Since $1/x\sqrt{1-x^2}$ is decreasing in $0 < x \leq 1/\sqrt{2}$, and $|\tau^{(m)}|T^2 \gg 1$, the alternating sign series here is bounded, and therefore $I \ll 1$. The lemma is proved. □

Lemma 5.4. *If $s > k$, then*

$$\int_{E_n} \prod_{i=1}^{n} \min\left(T, |\tau^{(i)}|^{-1/k}\right)^2 dx \ll |N(\alpha)|^{-1} T^{(s-k)n}.$$

Proof. Let

$$\alpha^{(\ell)}\zeta^{(\ell)} = u^{(\ell)}, \qquad \alpha^{(m)}\zeta^{(m)} = u_m e^{i\varphi_m} \tag{5.4}$$

The Jacobian of x_i with respect to u_ℓ, u_m, φ_m is

$$2^{r_2} D^{1/2} |N(\alpha)|^{-1} \prod_m u_m.$$

Therefore

$$\begin{aligned} \int_{E_n} \prod_{i=1}^{n} \min\left(T, |\tau^{(i)}|^{-1/k}\right)^2 dx \ll |N(\alpha)|^{-1} \prod_\ell \int_0^\infty \min(T^s, u_\ell^{-s/k})\, du_\ell \times \\ \times \prod_m \int_{-\pi}^{\pi} \int_0^\infty \min(T^{2s}, u_m^{-2s/k}) u_m\, du_m\, d\varphi_m. \end{aligned}$$

Since

$$\int_0^\infty \min(T^s, u^{-s/k})\, du \ll \int_0^{T^{-k}} T^s du + \int_{T^{-k}}^\infty u^{-s/k} du \ll T^{s-k} \tag{5.5}$$

and

$$\int_0^\infty \min(T^{2s}, u^{-2s/k}) u\, du = \frac{1}{2} \int_0^\infty \min(T^{2s}, v^{-s/k}) dv \ll T^{2(s-k)}, \tag{5.6}$$

the lemma follows. □

Remark. If $a > 1$, we have, similarly to (5.6),

$$\int_0^\infty \min\left(T^{2s}, a^{-2s/k}u^{-2s/k}\right) u\,du \ll a^{-2}\int_0^\infty \min\left(T^{2s}, v^{-s/k}\right)dv \ll a^{-2}T^{2(s-k)}.$$

5.3 Asymptotic Expansion for $S_i(\xi, T)$

We assume that $\xi \in B_\gamma$ $(\gamma \to \mathfrak{a})$ in a Farey division of U_n with respect to (h, t).

Lemma 5.5. *Let μ be an integer of K. Then*

$$\sum_{\substack{\lambda+\mu \in P(T) \\ \mathfrak{a}|\lambda}} E\big(\alpha_i(\lambda+\mu)^k\zeta\big) = N(\mathfrak{a})^{-1}I_i(\zeta, T) + O\big(N(\mathfrak{a})^{-1}\|\alpha_i\|h^{-1}T^{n+k-1}\big)$$

$$O\big(N(\mathfrak{a})^{-1}tT^{n-1}\big).$$

Proof. Let $\theta^{(i)}$ $(1 \le i \le n)$ be positive number with $\theta^{(m)} = \theta^{(m+r_2)}$ and determined by

$$\theta^{(i)}\max(h|\zeta^{(i)}|, t^{-1}) = D^{1/2n}\prod_{i=1}^{n}\max(h|\zeta^{(i)}|, t^{-1})^{1/n}N(\mathfrak{a})^{1/n}. \tag{5.7}$$

Then

$$\prod_{i=1}^{n}\theta^{(i)} = D^{1/2}N(\mathfrak{a}),$$

and it follows by Minkowski's theorem that there exists $\sigma \in \mathfrak{a}$ such that

$$0 < |\sigma^{(i)}| \le \theta^{(i)}, \qquad 1 \le i \le n. \tag{5.8}$$

Hence $\sigma\mathfrak{a}^{-1} = \mathfrak{b}$ is an integral ideal, and

$$N(\mathfrak{b}) = |N(\sigma)|\,N(\mathfrak{a})^{-1}.$$

Therefore $\mathfrak{b}$ belongs to a finite set depending on K only. Let $\sigma_1, \ldots, \sigma_n$ be a basis of $\mathfrak{b}^{-1}$. Then $\mathfrak{a} = \sigma\mathfrak{b}^{-1}$ has a basis

$$\tau_i = \sigma\sigma_i, \qquad 1 \le i \le n.$$

By (5.7) and (5.8) we obtain

$$\|\tau_i\| \ll \|\sigma\| \ll \|\theta\| \ll t.$$

Expressing λ in terms of τ_i $(1 \le i \le n)$ we have

$$\lambda = g_1\tau_1 + \cdots + g_n\tau_n,$$

where $g_i \in \mathbb{Z}$, $1 \le i \le n$. Let $U(\lambda)$ denote the unit cube

$$\{\mathbf{s} : \tau = s_1\tau_1 + \cdots + s_n\tau_n,\ g_i \le s_i < g_i + 1,\ 1 \le i \le n\}.$$

Then we have, by (5.7), $\|(\tau - \lambda)\zeta\| \ll \|\theta\zeta\| \ll h^{-1}$, and

$$\begin{aligned}\|\alpha_i(\tau+\mu)^k\zeta - \alpha_i(\lambda+\mu)^k\zeta\| &\ll \|\alpha_i\|\,\|(\tau-\lambda)\zeta\|\big(\|\tau+\mu\|^{k-1} + \|\lambda+\mu\|^{k-1}\big)\\ &\ll \|\alpha_i\|\,h^{-1}T^{k-1},\end{aligned}$$

and therefore

$$E\big(\alpha_i(\lambda+\mu)^k\zeta\big) = \int_{U(\lambda)} E\big(\alpha_i(\lambda+\mu)^k\zeta\big)ds + O\big(\|\alpha_i\|\,h^{-1}T^{k-1}\big),$$

where $ds = ds_1 \cdots ds_n$. It follows by Lemma 3.2 that the number of integers λ with $\mathfrak{a}|\lambda$ and $\lambda + \mu \in P(T)$ is $O\big(N(\mathfrak{a})^{-1}T^n\big)$. We have

$$\begin{aligned}\sum_{\substack{\lambda+\mu\in P(T)\\ \mathfrak{a}|\lambda}} E\big(\alpha_i(\lambda+\mu)^k\zeta\big) &= \sum_{\substack{\lambda+\mu\in P(T)\\ \mathfrak{a}|\lambda}} \int_{U(\lambda)} E\big(\alpha_i(\lambda+\mu)^k\zeta\big)ds\\ &\quad + O\big(N(\mathfrak{a})^{-1}\|\alpha_i\|h^{-1}T^{n+k-1}\big).\end{aligned}$$

Let F denote the domain in S-space defined by

$$0 \le \tau^{(\ell)} + \mu^{(\ell)} \le T, \qquad |\tau^{(m)} + \mu^{(m)}| \le T.$$

Then, by Lemma 3.2, the volume of the region belonging to exactly one of $\bigcup_{\lambda+\mu\in P(T),\, \mathfrak{a}|\lambda} U(\lambda)$ and F is dominated by $O\big(N(\mathfrak{a})^{-1}tT^{n-1}\big)$. Therefore we have

$$\begin{aligned}\sum_{\substack{\lambda+\mu\in P(T)\\ \mathfrak{a}|\lambda}} E\big(\alpha_i(\lambda+\mu)^k\zeta\big) &= \int_F E\big(\alpha_i(\lambda+\mu)^k\zeta\big)ds\\ &\quad O\big(N(\mathfrak{a})^{-1}\|\alpha_i\|h^{-1}T^{n+k-1}\big) + O\big(N(\mathfrak{a})^{-1}tT^{n-1}\big).\end{aligned}$$

Let $\tau + \mu = \eta$. Since the Jacobian of s_i with respect to y_i is

$$D^{1/2}|\det(\tau_j^{(i)})|^{-1} = N(\mathfrak{a})^{-1},$$

the lemma follows. □

Lemma 5.6. *We have*

$$\sum_{\lambda\in P(T)} E(\alpha_i\lambda^k\xi) = G_i(\gamma)I_i(\zeta, T) + O\big(\|\alpha_i\|h^{-1}T^{n+k-1}\big) + O\big(tT^{n-1}\big).$$

Proof. We have, by Lemma 5.5,

$$\sum_{\lambda \in P(T)} E(\alpha_i \lambda^k \xi) = \sum_{\mu(\mathfrak{a})} E(\alpha_i \mu^k \gamma) \sum_{\substack{\lambda+\mu \in P(T) \\ \mathfrak{a}|\lambda}} E(\alpha_i (\lambda+\mu)^k \zeta)$$

$$= \sum_{\mu(\mathfrak{a})} E(\alpha_i \mu^k \gamma) N(\mathfrak{a})^{-1} I_i(\zeta, T) + O(\|\alpha_i\| h^{-1} T^{n+k-1}) + O(t T^{n-1})$$

$$= G_i(\gamma) I_i(\zeta, T) + O(\|\alpha_i\| h^{-1} T^{n+k-1}) + O(t T^{n-1}),$$

since $\mathfrak{a}_i | \mathfrak{a}$. The lemma is proved. □

Remark. If (h,t) satisfies (5.2), then the error term in Lemma 5.6 is $O(\|\alpha_i\| T^{n-a})$.

5.4 Further Estimates on Basic Domains

We assume that (h,t) satisfies (5.2).

Lemma 5.7. *If $s \geq 4kn$, then*

$$\int_B S(\xi, T) E(-\nu\xi)\, dx = \sum_{\gamma \in \Gamma(t)} G(\gamma) E(-\nu\gamma) \int_{B_\gamma} I(\zeta, T) E(-\nu\zeta)\, dx$$
$$+ O\Big((A^s + A^{2sn/k}) T^{(s-k)n-a}\Big).$$

Proof. Since

$$1 \leq |N(\alpha_j)| = |\alpha_j^{(1)} \cdots \alpha_j^{(n)}| \leq |\alpha_j^{(i)}|\, A^{n-1},$$

we have

$$|\alpha_j^{(i)}|^{-1} \leq A^{n-1}, \qquad 1 \leq i \leq n, \quad 1 \leq j \leq s. \tag{5.9}$$

Let $\gamma \to \mathfrak{a}$. Since $\alpha_i \gamma \to \mathfrak{a}_i$, we have $\mathfrak{a}_i | \mathfrak{a} | \alpha_i \mathfrak{a}_i$ and

$$N(\mathfrak{a}) \leq |N(\alpha_i)|\, N(\mathfrak{a}_i).$$

Therefore

$$N(\mathfrak{a}_i)^{-1} \leq |N(\alpha_i)|\, N(\mathfrak{a})^{-1}. \tag{5.10}$$

By Theorem 2.1, Lemmas 5.3 and 5.6, and (5.10) we have

$$S(\xi, T) = G(\gamma) I(\zeta, T) + O\Big(A T^{n-a} \max_j \big(A T^{n-a}, |G_j(\gamma) I_j(\zeta, T)|\big)^{s-1}\Big)$$
$$= G(\gamma) I(\zeta, T) + O(A^s T^{(n-a)s})$$
$$+ \sum_{j=1}^{s} O\Big(A^{2sn/k} T^{n-a} N(\mathfrak{a})^{-\frac{s-1}{k}+\epsilon} \prod_{i=1}^{n} \min\big(T, |\alpha_j^{(i)} \zeta^{(i)}|^{-1/k}\big)^{s-1}\Big).$$

Since the number of $\gamma \in \Gamma(t)$ such that $\gamma \to \mathfrak{a}$ is $O\big(N(\mathfrak{a})\big)$, and the number of $\mathfrak{a}$ such that $N(\mathfrak{a}) = m$ is $O(m^{\epsilon})$ (see, for example, [Hua 1 §16.8]), we have

$$\sum_{\gamma\in\Gamma(t)} N(\mathfrak{a})^{-\frac{s-1}{k}+1+\epsilon} \ll \sum_{N(\mathfrak{a})\le t^n} N(\mathfrak{a})^{-\frac{s-1}{k}+\epsilon} \ll \sum_{m\le t^n} m^{-\frac{s-1}{k}+1+2\epsilon} \ll 1.$$

Therefore, by Lemma 5.4,

$$\begin{aligned}\int_B S(\xi,T)E(-\nu\xi)\,dx &= \sum_{\gamma\in\Gamma(t)} G(\gamma)E(-\nu\gamma)\int_{B_\gamma} I(\zeta,T)E(-\nu\zeta)dx + O\big(A^sT^{(n-a)s}\big)\\ &+O\Big(A^{2sn/k}T^{n-a}\sum_{\gamma\in\Gamma(t)} N(\mathfrak{a})^{-\frac{s-1}{k}+\epsilon}\sum_{j=1}^{s}\int_{E_n}\prod_{i=1}^{n}\min\Big(T,\,|\alpha_j^{(i)}\zeta^{(i)}|^{-1/k}\Big)^{s-1}dx\Big).\\ &= \sum_{\gamma\in\Gamma(t)} G(\gamma)E(-\nu\gamma)\int_{B_\gamma} I(\zeta,T)E(-\nu\zeta)dx + O\big(A^sT^{(n-a)s}\big) + O\big(A^{2sn/k}T^{(s-k)n-a}\big).\end{aligned}$$

Since $a = 1/4 + 1/4kn$, the lemma follows. □

Lemma 5.8. *If $s \ge 4kn$, then*

$$\int_B S(\xi,T)E(-\nu\xi)dx = \sum_{\gamma\in\Gamma(t)} G(\gamma)E(-\nu\gamma)\int_{E_n} I(\zeta,T)E(-\nu\zeta)dx + O\big(\Phi T^{(n-k)s}\big).$$

Proof. By Lemma 5.7, it suffices to show that

$$W = \sum_{\gamma\in\Gamma(t)} G(\gamma)E(-\nu\gamma)\int_{E_n-B_\gamma} I(\zeta,T)E(-\nu\zeta)\,dx + O\big(A^{2sn^2/k}T^{(s-k)n-(1-a)n}\big).$$

If $\mathbf{x} \in E_n - B_\gamma$, then there is at least one index i such that $h|\zeta^{(i)}| \ge N(\mathfrak{a})^{-1/n}$. Otherwise it follows from $t^{-1} \le N(\mathfrak{a})^{-1/n}$ that

$$\prod_{i=1}^{n}\max\big(h|\zeta^{(i)}|,\,t^{-1}\big) \le N(\mathfrak{a})^{-1},$$

that is $\mathbf{x} \in B_\gamma$, which leads to a contradiction. By Lemma 5.3 and (5.4), (5.9), we have

$$\begin{aligned}\int_{E_n-B_\gamma} I(\zeta,T)E(-\nu\zeta)dx &\ll \int_{E_n-B_\gamma}\prod_{j=1}^{s}\prod_{i=1}^{n}\min\big(T,|\alpha_j^{(i)}\zeta^{(i)}|^{-1/k}\big)\,dx\\ &\ll A^{sn^2/k}\int_{E_n-B_\gamma}\prod_{i=1}^{n}\min\big(T,|\zeta^{(i)}|^{-1/k}\big)^s\,dx,\end{aligned}$$

and

$$\int_{E_n - B_\gamma} \prod_{i=1}^{n} \min\left(T, |\zeta^{(i)}|^{-1/k}\right)^s dx$$
$$\ll \int_{h^{-1}N(\mathfrak{a})^{-1/n}}^{\infty} u^{-s/k} du \left(\int_0^\infty \min(T^s, v^{-s/k}) dv \right)^{r_1 - 1}$$
$$\times \left(\int_{-\pi}^{\pi} \int_0^\infty \min\left(T^{2s}, w^{-2s/k}\right) w \, dw \, d\varphi \right)^{r_2}$$

or

$$\ll \int_{-\pi}^{\pi} \int_{h^{-1}N(\mathfrak{a})^{-1/n}}^{\infty} u^{-2s/k+1} du \, d\varphi \left(\int_{-\pi}^{\pi} \int_0^\infty \min\left(T^{2s}, v^{-2s/k}\right) v \, dv \, d\theta \right)^{r_2 - 1}$$
$$\times \left(\int_0^\infty \min\left(T^s, w^{-s/k}\right) dw \right)^{r_1}$$

Therefore, by (5.5) and (5.6), we obtain

$$\int_{E_n - B_\gamma} I(\zeta, T) E(-\nu\zeta) dx \ll A^{sn^2/k} T^{(s-k)(n-1)} h^{s/k-1} N(\mathfrak{a})^{\frac{1}{n}(\frac{s}{k}-1)}$$

or

$$\ll A^{sn^2/k} T^{(s-k)(n-2)} h^{2s/k-2} N(\mathfrak{a})^{\frac{2}{n}(\frac{s}{k}-1)}.$$

By Theorem 2.1 and (5.10), we have

$$\sum_{\gamma \in \Gamma(t)} G(\gamma) N(\mathfrak{a})^{\frac{1}{n}(\frac{s}{k}-1)} \ll A^{sn/k} \sum_{\gamma \in \Gamma(t)} N(\mathfrak{a})^{\epsilon - \frac{s}{k} + \frac{s}{k} - 1}$$
$$\ll A^{sn/k} \sum_{N(\mathfrak{a}) \le t^n} N(\mathfrak{a})^{\epsilon} \ll A^{sn/k} t^{2n},$$

and

$$\sum_{\gamma \in \Gamma(t)} G(\gamma) N(\mathfrak{a})^{\frac{2}{n}(\frac{s}{k}-1)} \ll A^{sn/k} \sum_{\gamma \in \Gamma(t)} N(\mathfrak{a})^{\epsilon - \frac{s}{k} + \frac{s}{k} - 1} \ll A^{sn/k} t^{2n}.$$

Note that the latter case occurs only when $\mathbb{K}$ has complex conjugates, so that $n \ge 2$. Therefore

$$W \ll A^{2sn^2/k} T^{(s-k)n} \left(T^{-(\frac{s}{k}-1)(1-a)+2(1-a)n} + T^{-(\frac{2s}{k}-2)(1-a)+2(1-a)n} \right)$$
$$\ll A^{2sn^2/k} T^{(s-k)n-(1-a)n}.$$

The lemma is proved. □

5.5 Proof of Theorem 5.1

Let $\eta' = y_1'w_1 + \cdots + y_n'w_n$, $\zeta' = x_1'\rho_1 + \cdots + x_n'\rho_n$, $dy' = dy_1' \cdots dy_n'$, $dx' = dx_1' \cdots dx_n'$, $\eta = T\eta'$ and $\zeta = T^{-k}\zeta'$. The Jacobians of y_i and x_i with respect to y_i' and x_i' are T^n and T^{-kn} respectively. We have

$$\alpha_i\eta^k\zeta = \alpha_i{\eta'}^k\zeta', \qquad 1 \le i \le s.$$

Replacing η' and ζ' by η and ζ again, we find that $I(\zeta,T) = T^{sn}I(\zeta)$ and

$$\int_{E_n} I(\zeta,T)E(-\nu\zeta)\,dx = J(\mu)T^{(s-k)n} \qquad (\mu = \nu T^{-k}). \tag{5.11}$$

By Theorem 2.1 and (5.10),

$$\begin{aligned}\sum_{\substack{\gamma\to\mathfrak{a}\\ N(\mathfrak{a})>t^n}} G(\gamma)E(\nu\gamma) &\ll \sum_{N(\mathfrak{a})>t^n}\prod_{i=1}^{s}\left|N(\alpha_i)\right|^{1/k}N(\mathfrak{a})^{-s/k+1+\epsilon}\\ &\ll A^{sn/k}\sum_{N(\mathfrak{a})>t^n}N(\mathfrak{a})^{-s/k+1+\epsilon}\\ &\ll A^{sn/k}\sum_{m>t^n}m^{-s/k+1+2\epsilon} \ll A^{sn/k}T^{-(1-a)n},\end{aligned}$$

and thus

$$\sum_{\gamma\in\Gamma(t)} G(\gamma)E(-\nu\gamma) = \mathfrak{S}(\nu) + O\big(A^{sn/k}T^{-(1-a)n}\big).$$

Now we estimate the singular series for $s > k$. By Lemma 5.3 and (5.4), (5.9), we have

$$\begin{aligned}J(\nu) &\ll A^{sn^2/k}\prod_{\ell}\left(\int_0^1 \min\left(1, u_\ell^{-s/k}\right)du_\ell\right)\\ &\qquad\times\prod_{m}\left(\int_{-\pi}^{\pi}\int_0^1 \min\left(1, u_m^{-s/k}\right)u_m\,du_m\,d\varphi_m\right)\\ &\ll A^{sn^2/k}.\end{aligned}$$

Therefore, by Lemma 5.8,

$$\begin{aligned}\int_B S(\xi,T)E(-\nu\xi)dx &= \Big(\mathfrak{S}(\nu)+O(A^{sn/k}T^{-(1-a)n})\Big)J(\mu)T^{(s-k)n}+O(\Phi T^{(s-k)n})\\ &= \mathfrak{S}(\nu)J(\mu)T^{(s-k)n} + O\big(\Phi T^{(s-k)n}\big).\end{aligned}$$

The theorem is proved. □

Remark. The above argument shows that, if $s \ge 2k+1$, then

$$\mathfrak{S} \ll A^{sn/k}.$$

5.6 Proof of Theorem 5.2

Lemma 5.9 (Siegel). *Under the assumption of* Theorem 3.1, *we have* $\|\alpha\| > t$ *for* $\alpha^{-1}\beta = x_1\rho_1 + \cdots + x_n\rho_n$ *and* $\mathbf{x} \in S$.

Proof. By Theorem 3.1, we have α, β such that $1|\alpha$, $\delta^{-1}|\beta$, $0 < \|\alpha\| \le h$ and $\|\alpha\xi - \beta\| < h^{-1}$. Set $\alpha^{-1}\beta = \gamma$ and $\gamma\delta^{-1} = \mathfrak{b}/\mathfrak{a}$, $(\mathfrak{a}, \mathfrak{b}) = 1$; then $\mathfrak{a}|\alpha$ and thus $N(\mathfrak{a}) \le |N(\alpha)|$. Since $\mathbf{x} \in S$, we have

$$\prod_{i=1}^{n} \max\left(h|\xi^{(i)}\alpha^{(i)} - \beta^{(i)}|\,|\alpha^{(i)}|^{-1}, t^{-1}\right) = \prod_{i=1}^{n} \max\left(h|\xi^{(i)} - \gamma^{(i)}|, t^{-1}\right) > N(\mathfrak{a})^{-1}.$$

In fact, if we had

$$\prod_{i=1}^{n} \max\left(h|\xi^{(i)} - \gamma^{(i)}|, t^{-1}\right) \le N(\mathfrak{a})^{-1},$$

then $t^{-n} \le N(\mathfrak{a})^{-1}$, and so $\mathbf{x}$ would belong to basic domains, which is impossible. Therefore

$$\prod_{i=1}^{n} \max\left(|\alpha^{(i)}|^{-1}, t^{-1}\right) > N(\mathfrak{a})^{-1},$$

that is,

$$|N(\alpha)|^{-1} \prod_{i=1}^{n} \max\left(1, |\alpha^{(i)}|t^{-1}\right) > N(\mathfrak{a})^{-1}.$$

Consequently

$$\prod_{i=1}^{n} \max\left(1, |\alpha^{(i)}|t^{-1}\right) > |N(\alpha)|\, N(\mathfrak{a})^{-1} \ge 1,$$

and the lemma follows. □

Lemma 5.10. *Let* θ *be a number satisfying* $0 < \theta < (1-a)2^{1-k}$. *If* $\mathbf{x} \in S$, *then*

$$S_i(\xi, T) \ll T^{n-\theta}, \qquad 1 \le i \le s,$$

where the implied constant depends on θ.

Proof. By Lemma 5.9 and Theorem 3.2, we have

$$S_i(\xi, T) \ll T^{n+\epsilon-(1-a)2^{1-k}} \ll T^{n-\theta}$$

by taking a small ϵ depending on θ. The lemma follows. □

Lemma 5.11. *If $s \geq 2^k + 1$, then*

$$\int_S S(\xi, T)E(-\nu\xi)\,dx \ll T^{(s-k)n-\theta}.$$

Proof. Put $\theta_1 = \theta/2 + (1-a)2^{-k}$, so that $\theta < \theta_1 < (1-a)2^{1-k}$. By Theorem 5.1 and Lemma 5.10, we have

$$\int_S |S_i(\xi,T)|^s dx \ll T^{(s-2^k)(n-\theta_1)} \int_{U_n} |S_i(\xi,T)|^{2^k} dx$$
$$\ll T^{(s-2^k)(n-\theta_1)+(2^k-k)n+\theta_1-\theta} \ll T^{(s-k)n-\theta}.$$

Therefore, by Hölder's inequality,

$$\int_S S(\xi,T)E(-\nu\xi)\,dx \ll \int_S |S(\xi,T)|dx \ll \prod_{i=1}^{s} \left(\int_S |S(\xi,T)|^s dx \right)^{1/s} \ll T^{(s-k)n-\theta}.$$

The Lemma is proved. □

Set $\theta = 2^{-k}$. Then Theorem 5.2 follows immediately by Theorem 5.1 and Lemma 5.11. □

Notes

Siegel's generalised circle method includes the Farey division of U_n into B and S, and the evaluation of the integral over a basic domain described in this chapter was given in Siegel [3,4]. It was improved and simplified by many mathematicians; see, for example, the papers by Körner, Tatuzawa and Wang in Ref. II.

Lemma 5.9: See Siegel [3].

Chapter 6

Singular Series and Singular Integrals

6.1 Introduction

For a given $\wp$, let p be the rational prime contained in $\wp$, and let b, e, f denote the integers such that

$$\wp^e \| p, \quad p^b \| k \quad \text{and} \quad N(\wp) = p^f.$$

Define

$$t_0 = \left[\frac{e}{p-1}\right] + be + 1.$$

Let

$$H(\mathfrak{a}) = \sum_{\gamma}{}^{\star} G(\gamma) E(-\nu\gamma),$$

where γ runs over a reduced system of $(\mathfrak{a}\delta)^{-1} \bmod \delta^{-1}$. Then

$$\mathfrak{S}(\nu) = \sum_{\mathfrak{a}} H(\mathfrak{a}).$$

Theorem 6.1. *If there exists a positive integer $s_0 \geq 2k+1$ such that, for any $\wp$, the congruence*

$$\alpha_1 \lambda_1^k + \cdots + \alpha_s \lambda_s^k \equiv \nu \pmod{\wp^{t_0}} \tag{6.1}$$

has a solution with a λ_i such that $\wp \nmid \alpha_i \lambda_i$ provided that $s \geq s_0$, then

$$\mathfrak{S}(\nu) > c_1(s, A, K) > 0.$$

This theorem shows that the condition for the value of the singular series being positive is a purely arithmetical congruence condition.

Theorem 6.2. *If $s > k$, then the singular integral is absolutely convergent, and*

$$J(\mu) = D^{(1-s)/2} k^{-sn} |N(\alpha_1 \cdots \alpha_s)|^{-1/k} \prod_{\ell} F_\ell \prod_m H_m,$$

where

$$F_\ell = \int_{W_\ell} \prod_{i=1}^{s} w_i^{1/k-1} dw$$

with $dw = dw_1 \cdots dw_{s-1}$, *and* W_ℓ *is the domain*

$$0 \le w_i \le |\alpha_i^{(\ell)}|, \quad 1 \le i \le s, \quad \mu^{(\ell)} \pm w_1 \pm \cdots \pm w_s = 0.$$

Here the sign before w_i *is the sign of* $-\alpha_i^{(\ell)}$, $1 \le i \le s$, *and where*

$$H_m = \int_{V_m} \prod_{i=1}^{s} w_i^{1/k-1} dw\, d\varphi$$

with $d\varphi = d\varphi_1 \cdots d\varphi_{s-1}$, *and* V_m *is the domain*

$$0 \le w_i \le |\alpha_i^{(m)}|^2, \quad 1 \le i \le s, \qquad -\pi \le \varphi_j \le \pi, \quad 1 \le j \le s-1,$$
$$w_s = \left|\mu^{(m)} - w_1^{1/2} e^{i\varphi_1} - \cdots - w_{s-1}^{1/2} e^{i\varphi_{s-1}}\right|^2.$$

6.2 Product form for the Singular Series

Let

$$\chi(\wp) = \sum_{i=0}^{\infty} H(\wp^i).$$

By Theorem 2.1, we see that the series is absolutely convergent and is equal to

$$1 + O\left(N(\wp)^{-s/k+1+\epsilon}\right)$$

if $s \ge 2k+1$.

Lemma 6.1. *If* $s \ge 2k+1$, *then*

$$\mathfrak{S}(\nu) = \prod_{\wp} \chi(\wp).$$

To prove Lemma 6.1 we shall need:

Lemma 6.2. *Suppose that* $\gamma_j \to \mathfrak{a}_j$, $j = 1, 2$, *and* $(\mathfrak{a}_1, \mathfrak{a}_2) = 1$. *Then*

$$\gamma_1 + \gamma_2 \to \mathfrak{a}_1 \mathfrak{a}_2$$

and

$$G_i(\gamma_1 + \gamma_2) = G_i(\gamma_1) G_i(\gamma_2).$$

Proof. Let $\gamma_j\delta = \mathfrak{b}_j/\mathfrak{a}_j$, $(\mathfrak{a}_j, \mathfrak{b}_j) = 1$, $j = 1, 2$. Since $(\mathfrak{a}_2\mathfrak{b}_1 + \mathfrak{a}_1\mathfrak{b}_2, \mathfrak{a}_1\mathfrak{a}_2) = 1$ and

$$(\gamma_1 + \gamma_2)\delta = \frac{\mathfrak{b}_1}{\mathfrak{a}_1} + \frac{\mathfrak{b}_2}{\mathfrak{a}_2} = \frac{\mathfrak{a}_2\mathfrak{b}_1 + \mathfrak{a}_1\mathfrak{b}_2}{\mathfrak{a}_1\mathfrak{a}_2},$$

we have the first assertion of the lemma.

Now let $\alpha_i\gamma_j\delta = \mathfrak{b}_{ij}/\mathfrak{a}_{ij}$, $(\mathfrak{a}_{ij}, \mathfrak{b}_{ij}) = 1$, $j = 1, 2$. Then $\mathfrak{a}_{ij}|\mathfrak{a}_j$ and

$$G_i(\gamma_j) = N(\mathfrak{a}_j)^{-1} \sum_{\lambda(\mathfrak{a}_j)} E(\alpha_i\lambda^k\gamma_j).$$

We can take two integers β_1 and β_2 such that $(\beta_j, \mathfrak{a}_1\mathfrak{a}_2) = \mathfrak{a}_j$, $j = 1, 2$. Put $\lambda = \beta_1\lambda_2 + \beta_2\lambda_1$. Then if λ_1 and λ_2 run over complete systems $\bmod \mathfrak{a}_1$ and $\bmod \mathfrak{a}_2$ respectively, then λ runs over a complete residue system $\bmod \mathfrak{a}_1\mathfrak{a}_2$; see Lemma 2.4. We have

$$\begin{aligned}\sum_{\lambda_1(\mathfrak{a}_1)} E(\alpha_i\lambda_1^k\gamma_1) \sum_{\lambda_2(\mathfrak{a}_2)} E(\alpha_i\lambda_2^k\gamma_2) &= \sum_{\lambda_1(\mathfrak{a}_1)} \sum_{\lambda_2(\mathfrak{a}_2)} E\big(\alpha_i(\beta_2\lambda_1)^k\gamma_1 + \alpha_i(\beta_1\lambda_2)^k\gamma_2\big) \\ &= \sum_{\lambda_1(\mathfrak{a}_1)} \sum_{\lambda_2(\mathfrak{a}_2)} E\big(\alpha_i(\beta_2\lambda_1 + \beta_1\lambda_2)^k(\gamma_1 + \gamma_2)\big).\end{aligned}$$

The second assertion is proved. □

Lemma 6.3. *Suppose that* $(\mathfrak{a}_1, \mathfrak{a}_2) = 1$. *Then*

$$H(\mathfrak{a}_1, \mathfrak{a}_2) = H(\mathfrak{a}_1)H(\mathfrak{a}_2).$$

Proof. By Lemma 6.2,

$$\begin{aligned}H(\mathfrak{a}_1)H(\mathfrak{a}_2) &= \sum_{\gamma_1}{}^{\star} G(\gamma_1)E(-\nu\gamma_1) \sum_{\gamma_2}{}^{\star} G(\gamma_2)E(-\nu\gamma_2) \\ &= \sum_{\gamma_1}{}^{\star} \sum_{\gamma_2}{}^{\star} G(\gamma_1 + \gamma_2)E\big(-\nu(\gamma_1 + \gamma_2)\big).\end{aligned}$$

If γ_j runs over a reduced residue system of $(\mathfrak{a}_j\delta)^{-1} \bmod \delta^{-1}$, $j = 1, 2$, then $\gamma_1 + \gamma_2$ runs over a reduced residue system of $(\mathfrak{a}_1\mathfrak{a}_2\delta)^{-1} \bmod \delta^{-1}$. The lemma is proved. □

Proof of Lemma 6.1. By Lemma 6.3,

$$\prod_{N(\wp)\le x} \chi(\wp) = \prod_{N(\wp)\le x} \Big(\sum_{i=0}^{\infty} H(\wp^i)\Big) = \sum_{N(\mathfrak{a})\le x} H(\mathfrak{a}) + \sum_{N(\mathfrak{a})>x}{}' H(\mathfrak{a}),$$

where Σ' means that $\mathfrak{a}$ runs over all ideals with prime ideal divisors having norm not exceeding x. Since the singular series is absolutely convergent if $s \ge 2k+1$, we have

$$\sum_{N(\wp)>x}{}' H(\mathfrak{a}) = o(1) \qquad \text{as} \quad x \to \infty.$$

The lemma is proved. □

6.3 Singular Series and Congruences

Let $M(\mathfrak{a}) = M(\nu, \mathfrak{a})$ denote the number of solutions of the congruence

$$\alpha_1 \lambda_1^k + \cdots + \alpha_s \lambda_s^k \equiv \nu \pmod{\mathfrak{a}},$$

where $\lambda_i \, (1 \le i \le s)$ run independently through complete systems of residues to the modulus $\mathfrak{a}$.

Lemma 6.4. *If there exists positive integers $s_0 \ge 2k+1$ such that the relation $M(\wp^t) \ge N(\wp)^{(t-t_0)(s-1)}$ holds for any $t \ge t_0$, $s \ge s_0$ and $\wp$, then $\mathfrak{S}(\nu) > c_2(s, A, K) > 0$.*

To prove Lemma 6.4 we shall need:

Lemma 6.5.

$$M(\mathfrak{a}) = N(\mathfrak{a})^{s-1} \sum_{\mathfrak{b}|\mathfrak{a}} H(\mathfrak{b}).$$

Proof. By Lemma 2.5,

$$M(\mathfrak{a})N(\mathfrak{a}) = \sum_{\gamma} \sum_{\lambda_1(\mathfrak{a})} \cdots \sum_{\lambda_s(\mathfrak{a})} E\big((\alpha_1 \lambda_1^k + \cdot + \alpha_s \lambda_s^k - \nu)\gamma\big),$$

where γ runs over a complete residue system of $(\mathfrak{a}\delta)^{-1} \bmod \delta^{-1}$. For $\mathfrak{b}|\mathfrak{a}$, let $\Sigma^\star_{\gamma_1}$ denote a sum with γ_1 over a reduced system of $(\mathfrak{b}\delta)^{-1} \bmod \delta^{-1}$. Then

$$\begin{aligned} M(\mathfrak{a})N(\mathfrak{a}) &= \sum_{\mathfrak{b}|\mathfrak{a}} \left(\frac{N(\mathfrak{a})}{N(\mathfrak{b})}\right)^s \sum_{\gamma_1}{}^{\star} \sum_{\lambda_1(\mathfrak{b})} \cdots \sum_{\lambda_s(\mathfrak{b})} E\big((\alpha_1 \lambda_1^k + \cdot + \alpha_s \lambda_s^k - \nu)\gamma_1\big) \\ &= N(\mathfrak{a})^s \sum_{\mathfrak{b}|\mathfrak{a}} \sum_{\gamma_1}{}^{\star} G(\gamma_1) E(-\nu\gamma_1) = N(\mathfrak{a})^s \sum_{\mathfrak{b}|\mathfrak{a}} H(\mathfrak{b}). \end{aligned}$$

The lemma is proved. □

Set $\mathfrak{a} = \wp^t$. Then

$$\sum_{i=0}^{t} H(\wp^i) = N(\wp)^{-t(s-1)} M(\wp^t). \tag{6.2}$$

Proof of Lemma 6.4. By (6.2), we have, for $t \geq t_0$,

$$\sum_{i=0}^{t} H(\wp^i) \geq N(\wp)^{-t(s-1)+(t-t_0)(s-1)} = N(\wp)^{-t_0(s-1)}.$$

Letting $t \to \infty$ we have $\chi(\wp) \geq N(\wp)^{-t_0(s-1)}$. By (5.10) and Theorem 2.1 with $\epsilon = 1/4ks$, it follows that if $N(\wp) \geq c_3(s, A, K)$, then

$$\sum_{i=0}^{\infty} H(\wp^i) \ll A^{sn/k} \sum_{i=1}^{\infty} N(\wp)^{i(1-s/k+1/4k)} < N(\wp)^{-1-1/2k},$$

and so

$$\chi(\wp) > 1 - N(\wp)^{-1-1/2k}.$$

Therefore

$$\begin{aligned} \mathfrak{S}(\nu) &= \prod_{N(\wp)\leq c_3} \chi(\wp) \prod_{N(\wp)>c_3} \chi(\wp) \\ &\geq \prod_{N(\wp)\leq c_3} N(\wp)^{-t_0(s-1)} \prod_{N(\wp)>c_3} \left(1 - N(\wp)^{-1-1/2k}\right) = c_2 > 0. \end{aligned}$$

The lemma is proved. □

6.4 $\wp$-adic Valuation

The non-Archimedean valuation of a field $\mathbb{K}$ is similar to that of the rational field $\mathbb{Q}$; see, for example, [Hua 1, Chapter 15]. For $\alpha \in \mathbb{K}$, let $\omega(\alpha) = \omega_\wp(\alpha)$ be the exponent with which $\wp$ enters into $\alpha\delta$. Let $\mathbb{K}_\wp$ be the completion of $\mathbb{K}$ with respect to $\wp$-adic valuation. Suppose that A is a number of $\mathbb{K}_\wp$ and is defined by a Cauchy sequence $\{\alpha_n\}$, $\alpha_n \in \mathbb{K}$. Since $\lim_{n\to\infty} \omega(\alpha_n)$ exists, we may denote it by $\omega(A)$. Then

$$\omega(AB) = \omega(A)\omega(B) \tag{6.3}$$

and

$$\omega(A+B) \geq \min\left(\omega(A), \omega(B)\right) \tag{6.4}$$

for every A, B in $\mathbb{K}_\wp$, where equality in (6.4) holds if $\omega(A) \neq \omega(B)$.

The series $\sum_{i=1}^{\infty} A_i$ converges in $\mathbb{K}_\wp$ if and only if $\omega(A_n) \to \infty$ as $n \to \infty$. In fact, if we set $S_t = \sum_{1\leq i\leq t} A_i$, then the series converges if and only if $\{S_t\}$ is a Cauchy sequence. If $\omega(A_n) \to \infty$, then by (6.4), $\omega(S_{t+q} - S_t) \geq \min\left(\omega(A_{t+1}), \ldots, \omega(A_{t+q})\right) \to \infty$, as $t \to \infty$. On the other hand, we have $\omega(A_t) = \omega(S_t - S_{t-1}) \to \infty$. The assertion follows.

Let $f(x)$, $g(x)$ and $h(x)$ be power series in $\mathbb{K}_\wp$, formally satisfying $f(x)g(x) = h(x)$. By (6.3) and (6.4), it follows that if $f(A)$ and $g(A)$ are convergent, then $h(A)$ is also convergent and satisfies $f(A)g(A) = h(A)$.

We define as usual the $\omega_p(m)$ for $m \in \mathbb{Q}$. Let

$$m = a_0 + a_1 p + \cdots + a_r p^r, \quad 0 \le a_i < p, \quad 1 \le i \le r,$$

be a p-adic representation of a rational integer m, and define

$$S(m) = a_0 + a_1 + \cdots + a_r.$$

Then $S(m) \ge 1$ if $m \ge 1$. Since

$$\begin{aligned}\omega_p(m!) &= \left[\frac{m}{p}\right] + \left[\frac{m}{p^2}\right] + \cdots + \left[\frac{m}{p^r}\right] \\ &= (a_1 + \cdots + a_r p^{r-1}) + (a_2 + \cdots + a_r p^{r-2}) + \cdots + a_r \\ &= a_1 + a_2(p+1) + \cdots + a_r(p^{r-1} + \cdots + 1) = \frac{m - S(m)}{p-1},\end{aligned}$$

we have, by (6.3),

$$\omega\left(\frac{A^m}{m!}\right) = \omega(A) + (m-1)\left(\omega(A) - \frac{e}{p-1}\right) + \frac{S(m)-1}{p-1}e.$$

Therefore the series

$$1 + A + \cdots + \frac{1}{m!}A^m + \cdots$$

is convergent, provided that

$$\omega(A) > \frac{e}{p-1};$$

we denote the sum by $\exp A$. Since $\log x/(x-1)$ is decreasing for $x \ge 2$, we have

$$\begin{aligned}\omega\left(\frac{B^m}{m}\right) &= m\,\omega(B) - e\,\omega(m) \ge m\,\omega(B) - \frac{\log m}{\log p}e \\ &\ge m\,\omega(B) - \frac{m-1}{p-1}e = \omega(B) + (m-1)\left(\omega(B) - \frac{e}{p-1}\right), \quad m \ge p.\end{aligned}$$

Therefore the series

$$B - \frac{1}{2}B^2 + \cdots + (-1)^{m-1}\frac{B^m}{m} + \cdots$$

is convergent, provided that

$$\omega(B) > \frac{e}{p-1};$$

we denote its sum by $\log(1+B)$.

Furthermore, if $\omega(A) > e/(p-1)$ and $\omega(B) > e/(p-1)$, then $\log(\exp A) = A$ and $\exp\big(\log(1+B)\big) = 1+B$. In fact, on setting $B = \exp A - 1$, we have

$$B - \frac{1}{2}B^2 + \cdots + (-1)^{m-1}\frac{B^m}{m} = A + \sum_{t>m} \sum_{\substack{i_1+\cdots+i_j=t \\ 1\le j\le m}} \left(\pm\frac{1}{j}\cdot\frac{A^{i_1+\cdots+i_j}}{i_1!\cdots i_j!}\right).$$

Since

$$\omega\left(\frac{1}{j}\cdot\frac{A^{i_1+\cdots+i_j}}{i_1!\cdots i_j!}\right) \ge (i_1+\cdots+i_j)\omega(A) - \frac{i_1-S(i_1)}{p-1}e - \cdots - \frac{i_j-S(i_j)}{p-1}e - \frac{j-1}{p-1}e$$
$$\ge t\left(\omega(A) - \frac{e}{p-1}\right) + \frac{e}{p-1},$$

the assertion $\log(\exp A) = A$ follows by letting $m \to \infty$. Set $A = \log(1+B)$. Then

$$1 + A + \cdots + \frac{A^m}{m!} = 1 + B + \sum_{\substack{i_1+\cdots+i_j=t \\ 1\le j\le m}} \left(\pm\frac{1}{j}\cdot\frac{A^{i_1+\cdots+i_j}}{i_1\cdots i_j}\right)$$

and

$$\omega\left(\frac{1}{j}\cdot\frac{B^{i_1+\cdots+i_j}}{i_1\cdots i_j}\right) \ge (i_1+\cdots+i_j)\omega(B) - \frac{i_1-1}{p-1}e - \cdots - \frac{i_j-1}{p-1}e - \frac{j-S(j)}{p-1}e$$
$$\ge t\left(\omega(B) - \frac{e}{p-1}\right) + \frac{e}{p-1}.$$

Letting $m \to \infty$, we have $\exp\big(\log(1+B)\big) = 1+B$.

If $\omega(A_i) > e/(p-1)$, $i = 1, 2$, then, by the product of two convergent series, we have

$$\exp(A_1 + A_2) = \exp A_1 \cdot \exp A_2.$$

If $\omega(B_i - 1) > e/(p-1)$, $i = 1, 2$, then

$$\exp(\log B_1 + \log B_2) = \exp(\log B_1)\cdot\exp(\log B_2) = B_1B_2,$$

and therefore

$$\log B_1B_2 = \log\big(\exp(\log B_1 + \log B_2)\big) = \log B_1 + \log B_2.$$

6.5 k-th Power Residues

Lemma 6.6. *Suppose that $\wp \nmid \beta$. If the congruence*

$$\beta\xi^k \equiv \alpha \pmod{\wp^{t_0}}$$

has a nontrivial solution, that is a solution satisfying $\wp\nmid\xi$, then so does the congruence

$$\beta\xi^k \equiv \alpha \pmod{\wp^{t}}$$

for any $t > t_0$.

Proof. Since $\wp \nmid \beta$, there exists $\bar{\beta}$ such that

$$\beta\bar{\beta} \equiv 1 \pmod{\wp^t}.$$

Set $\gamma = \alpha\bar{\beta}$. Then the congruences reduce to

$$\xi^k \equiv \gamma \pmod{\wp^{t_0}} \tag{6.5}$$

and

$$\xi^k \equiv \gamma \pmod{\wp^t}$$

respectively. If (6.5) is soluble nontrivially, then

$$(\gamma - \xi^k) = \wp^{t_0}\mathfrak{b},$$

where $\mathfrak{b}$ is an integral ideal. Since $\gamma - \xi^k = \xi^k(\gamma/\xi^k - 1)$ and $\wp \nmid \xi$, we have $\wp^{t_0} | (\gamma/\xi^k - 1)$, that is $\omega(\gamma/\xi^k - 1) \geq t_0 > e/(p-1)$, and we may write

$$\log\left(1 + (\frac{\gamma}{\xi^k} - 1)\right) = \log\frac{\gamma}{\xi^k}.$$

Since

$$\omega\left(\frac{1}{k}\log\frac{\gamma}{\xi^k}\right) = \omega\left(\log\frac{\gamma}{\xi^k}\right) - \omega(k) = \omega\left(\frac{\gamma}{\xi^k} - 1\right) - \omega(k) \geq t_0 - be > \frac{e}{p-1},$$

we may also write

$$\exp\left(\frac{1}{k}\log\frac{\gamma}{\xi^k}\right) = B,$$

say. Then

$$\log\left(\exp\left(\frac{1}{k}\log\frac{\gamma}{\xi^k}\right)\right) = \log B, \quad \log\frac{\gamma}{\xi^k} = \log B^k \quad \text{and} \quad \gamma = (\xi B)^k.$$

Therefore ξB is an integer. Choose the integer η so that $\omega(\xi B - \eta) \geq t$. Then

$$\omega(\eta^k - \gamma) = \omega(\eta^k - (\xi B)^k) \geq t,$$

and so

$$\eta^k \equiv \gamma \pmod{\wp^t}.$$

The lemma is proved. □

6.6 Proof of Theorem 6.1

Theorem 6.1 follows by Lemma 6.4 and the following:

Lemma 6.7. *If there exists an integer $s_0 \geq 2k+1$ such that the congruence (6.1) has a solution with $\wp \nmid \alpha_i\lambda_i$ provided that $s \geq s_0$, then, for any $t \geq t_0$,*

$$M(\wp^t) \geq N(\wp)^{(t-t_0)(s-1)}. \tag{6.6}$$

Proof. Suppose that $i = 1$ and that λ is a solution of (6.1) with $\wp \nmid \alpha_1\lambda_1$. Let π be an integer such that $\wp \| \pi$, and let

$$\mu_i = \lambda_i + \pi^{t_0}\nu, \qquad 2 \le i \le s,$$

where ν runs over a complete residue system mod $\wp^{t-t_0}$. It follows by Lemma 6.6 that

$$\alpha_1\lambda^k \equiv -\alpha_2\mu_2^k - \cdots - \alpha_s\mu_s^k \pmod{\wp^t}$$

always has a solution λ for any given $\mu_2, \ldots, \mu_s$. In other words (6.6) holds, and the lemma is proved. □

6.7 Monotonic Functions

If $F(\mathbf{x}) = F(x_1, \ldots, x_s)$ is nondecreasing for the variables $x_{i_1}, \ldots, x_{i_j}$ and non-increasing for the other variables over the rectangle

$$I = \{\mathbf{x} : 0 \le x_i \le c_i,\ 1 \le i \le s\},$$

then $F(\mathbf{x})$ is said to be monotonic over I.

Lemma 6.8 (Tatuzawa). *Let $F(\mathbf{x})$ be a finite product of bounded monotonic functions over I. If we write*

$$\chi_\lambda(x) = \frac{\sin 2\pi\lambda x}{\pi x},$$

then

$$\lim_{\lambda_i \to \infty} \int_I F(\mathbf{x}) \prod_{i=1}^{s} \chi_{\lambda_i}(x_i)\, dx = 2^{-s} F(+0).$$

To prove this we shall need:

Lemma 6.9. *Let $F(\mathbf{x})$ be a bounded monotonic function over I. Then $F(\mathbf{x})$ is summable (measurable and absolutely integrable) over I.*

Proof. We consider the case $s = 2$ only, since the case $s > 2$ can be treated similarly. We assume that $F(x_1, x_2)$ is nondecreasing for x_1 and x_2 over

$$I = \{(x_1, x_2) : 0 \le x_1 \le c_1, 0 \le x_2 \le c_2\}.$$

Otherwise if (for example) $F(x_1, x_2)$ is nonincreasing for x_1 and nondecreasing for x_2, we may consider the function $F(c_1 - x_1, x_2)$ instead of $F(x_1, x_2)$. From the points $(\alpha, 0)\,(0 \le \alpha \le c_1)$ and $(0, \beta)\,(0 \le \beta \le c_2)$, we draw lines parallel to the diagonal of I, joining $(0,0)$ to (c_1, c_2). Let γ be any given number in $\mathbb{R}$. On one of these lines we take a point p_α or a point p_β with the smallest coordinates (r,s) such that

$$F(x_1,x_2) \geq \gamma \qquad \text{whenever} \quad x_1 \geq r,\ x_2 \geq s,$$

except for a set of measure zero consisting of points lying upon the lines $x_1 = r$ and $x_2 = s$. We denote by Q_α or R_β the set

$$\{(x_1,x_2):\ F(x_1,x_2) \geq \gamma,\ x_1 \geq r,\ x_2 \geq s\},$$

and by S the set

$$\{(x_1,x_2):\ (x_1,x_2) \in I,\ F(x_1,x_2) \geq \gamma\}.$$

Clearly

$$S = \bigcup Q_\alpha \cup \bigcup R_\beta.$$

Take rational numbers a, b from the intervals $[0,c_1]$ and $[0,c_2]$ respectively and set

$$S = \bigcup Q_a \cup \bigcup R_b \cup E.$$

For any $\epsilon > 0$, we may make $\bar{m}E < \epsilon$ by taking sufficiently many a, b. Consequently we can infer that

$$S = \bigcup Q_a \cup \bigcup R_b \cup E_0, \qquad mE_0 = 0,$$

where a and b run over all rational numbers lying in the intervals cited above. Therefore S is a measurable set, and the lemma is proved. □

Proof of Lemma 6.8. Again we shall only deal with the case $s = 2$, and assume that $F(x_1,x_2)$ is a monotonic function over I. We assume that it is nondecreasing for x_1 and nonincreasing for x_2 over I. Since for $c > 0$,

$$\lim_{\lambda\to\infty} \int_0^c \chi_\lambda(x)\,dx = \lim_{\lambda\to\infty} \int_0^{2\pi\lambda c} \frac{\sin y}{y}\,dy = \frac{1}{\pi}\int_0^\infty \frac{\sin y}{y}\,dy = \frac{1}{2},$$

we may assume that $F(+0,+0) = 0$. Let the integral be divided as follows:

$$\int_0^{c_1}\int_0^{c_2} = \int_0^{a_1}\int_0^{a_2} + \int_{a_1}^{c_1}\int_0^{c_2} + \int_0^{c_1}\int_{a_2}^{c_2} - \int_{a_1}^{c_1}\int_{a_2}^{c_2},$$

where a_1 and a_2 are taken so as to satisfy

$$0 \leq |F(a_1,a_2)| < \epsilon, \qquad 0 < a_1 < c_1,\ 0 < a_2 < c_2.$$

By the second mean value theorem, we obtain

$$\int_0^{a_1}\int_0^{a_1} F(x_1,x_2)\chi_{\lambda_1}(x_1)\chi_{\lambda_2}(x_2)\,dx_1\,dx_2$$
$$= \int_0^{a_1} F(x_1,+0)\chi_{\lambda_1}(x_1)\,dx_1 \int_0^{\eta} \chi_{\lambda_2}(x_2)\,dx_2$$
$$= F(a_1-0,+0)\int_{\xi}^{a_1} \chi_{\lambda_1}(x_1)\,dx_1 \int_0^{\eta} \chi_{\lambda_2}(x_2)\,dx_2,$$

where $0 \le \xi \le a_1$, $0 \le \eta \le a_2$. Therefore

$$\int_0^{a_1}\int_0^{a_2} = O(\epsilon),$$

and the other three integrals in the division can be made arbitrarily small in absolute values by taking λ_1 and λ_2 sufficiently large. The lemma therefore holds.

We now proceed to prove the lemma when F is a finite product of bounded monotonic functions. We may assume that F is only a product of two monotonic functions F_1 and F_2, where F_1 is nondecreasing for x_1 and nonincreasing for x_2, and F_2 is contrariwise. We assume further that

$$0 \le |F_i(x_1,x_2)| \le A, \qquad i=1,2,$$

in I. We take a_1, a_2 such that

$$0 \le |F_i(a_1,a_2)| < \epsilon, \qquad 0 < a_i < c_i, \quad i=1,2.$$

Then, by the second mean value theorem, we obtain

$$\int_0^{a_1}\int_0^{a_1} F(x_1,x_2)\chi_{\lambda_1}(x_1)\chi_{\lambda_2}(x_2)\,dx_1\,dx_2$$
$$= \int_0^{a_1}\int_0^{a_1} AF_2(x_1,x_2)\chi_{\lambda_1}(x_1)\chi_{\lambda_2}(x_2)\,dx_1\,dx_2$$
$$- \int_0^{a_1}\int_0^{a_1} (A-F_1(x_1,x_2))F_2(x_1,x_2)\chi_{\lambda_1}(x_1)\chi_{\lambda_2}(x_2)\,dx_1\,dx_2$$
$$= \int_0^{a_1} AF_2(x_1,a_2-0)\chi_{\lambda_1}(x_1)\,dx_1 \int_{\eta_1}^{a_2} \chi_{\lambda_2}(x_2)\,dx_2$$
$$- \int_0^{a_1} (A-F_1(x_1,a_2-0))F_2(x_1,a_2-0)\chi_{\lambda_1}(x_1)\,dx_1 \int_{\eta_2}^{a_2} \chi_{\lambda_2}(x_2)\,dx_2$$
$$= AF_2(+0,a_2-0)\int_0^{\xi_1} \chi_{\lambda_1}(x_1)\,dx_1 \int_{\eta_1}^{a_2} \chi_{\lambda_2}(x_2)\,dx_2$$
$$-(A-F_1(+0,a_2-0))F_2(+0,a_2-0)\int_0^{\xi_2} \chi_{\lambda_1}(x_1)\,dx_1 \int_{\eta_2}^{a_2} \chi_{\lambda_2}(x_2)\,dx_2$$
$$= O(\epsilon),$$

where $0 \le \xi_i \le a_1$, $0 \le \eta_i \le a_2$, $i = 1, 2$. The remaining part of the integral is arbitrarily small in absolute value if λ_1 and λ_2 are sufficiently large. The lemma is proved. □

6.8 Proof of Theorem 6.2

The singular integral is absolutely convergent if $s > k$ (see §5.5). Therefore

$$J(\mu) = \lim_{\lambda_\ell, \lambda_m, \lambda'_m \to \infty} J(\Omega), \qquad J(\Omega) = \int_\Omega I(\zeta) E(-\mu\zeta)\, dx,$$

where Ω denotes the closed region $\mathbf{x}$ defined by

$$|v_\ell| \le \lambda_\ell, \qquad |v_m| \le \lambda_m, \qquad |v'_m| \le \lambda'_m,$$

in which

$$v_\ell = \zeta^{(\ell)}, \qquad v_m = \frac{\zeta^{(m)} + \zeta^{(m+r_2)}}{\sqrt{2}}, \qquad v'_m = \frac{\zeta^{(m)} - \zeta^{(m+r_2)}}{\sqrt{2}i}.$$

Consider the sn variables y_{ij}, $1 \le i \le s$, $1 \le j \le n$. Let

$$\eta_i = y_{i1} w_1 + \cdots + y_{in} w_n, \qquad dY_i = dy_{i1} \cdots dy_{in}$$

and P_i be the domain:

$$0 \le \eta^{(\ell)} \le 1, \qquad |\eta^{(m)}| \le 1.$$

Let

$$\alpha_i^{(i)} \eta_1^{(i)k} + \cdots + \alpha_s^{(i)} \eta_s^{(i)k} - \mu^{(i)} = z_i, \qquad 1 \le i \le n,$$

and

$$u_\ell = z_\ell, \qquad u_m = \frac{z_m + z_{m+r_2}}{\sqrt{2}}, \qquad u'_m = -\frac{z_m - z_{m+r_2}}{\sqrt{2}i}.$$

Since

$$\zeta^{(m)} = \frac{v_m + iv'_m}{\sqrt{2}}, \qquad \zeta^{(m+r_2)} = \frac{v_m - iv'_m}{\sqrt{2}},$$

$$z_m = \frac{u_m - iu'_m}{\sqrt{2}}, \qquad z_{m+r_2} = \frac{u_m + iu'_m}{\sqrt{2}},$$

we have

$$\sum_{i=1}^n \zeta^{(i)} z_i = \sum_\ell u_\ell v_\ell + \sum_m (u_m v_m + u'_m v'_m).$$

The Jacobian of x_i with respect to v_ℓ, v_m, v'_m is equal to

$$\left| \det(\rho_j^{(i)}) \right|^{-1} |-i|^{r_2} = D^{1/2}.$$

Set $dv = \prod_\ell dv_\ell \prod_m dv_m dv'_m$. We then have

$$\begin{aligned}J(\Omega) &= \int_{P_1}\cdots\int_{P_s} dY_1\cdots dY_s \int_\Omega e\Big(\sum_{i=1}^n \zeta^{(i)} z_i\Big)\,dx \\ &= D^{1/2}\int_{P_1}\cdots\int_{P_s} dY_1\cdots dY_s\times \\ &\quad\times\int_\Omega e\Big(\sum_\ell u_\ell v_\ell + \sum_m (u_m v_m + u'_m v'_m)\zeta^{(i)} z_i\Big)\,dv \\ &= D^{1/2}\int_{P_1}\cdots\int_{P_s}\prod_\ell \chi_{\lambda_\ell}(u_\ell)\prod_m \chi_{\lambda_m}(u_m)\chi_{\lambda'_m}(u'_m)dY_1\cdots dY_s.\end{aligned}$$

Let

$$z_i = t_1 w_1^{(i)} + \cdots + t_n w_n^{(i)}, \qquad 1\le i\le n.$$

Then

$$\alpha_s^{(i)}\eta_s^{(i)k} = z_i - \Big(\alpha_1^{(i)}\eta_1^{(i)k} + \cdots + \alpha_{s-1}^{(i)}\eta_{s-1}^{(i)k} - \mu^{(i)}\Big).$$

The Jacobian of y_{si} with respect to t_i is equal to

$$\Big|\det\big(k\alpha_s^{(i)}\eta_s^{(i)k-1}w_j^{(i)}\big)\Big|^{-1}\,|\det(w_j^{(i)}| = N\big(k^{-1}|\alpha_s^{-1}\eta_s^{1-k}|\big)$$

and the Jacobian of t_i with respect to u_ℓ, u_m, u'_m is

$$\big|\det(w_j^{(i)})\big|^{-1}|i|^{r_2} = D^{-1/2}.$$

Therefore

$$\begin{aligned}J(\Omega) &= \int_Q \prod_\ell \chi_{\lambda_\ell}(u_\ell)\prod_m \chi_{\lambda_m}(u_m)\chi_{\lambda'_m}(u'_m)\,du\times \\ &\quad\times\int_{P_1}\cdots\int_{P_{s-1}} N\big(k^{-1}|\alpha_s^{-1}\eta_s^{1-k}|\big)dY_1\cdots dY_{s-1},\end{aligned}$$

where Q is a closed region containing the origin of u in its interior, and $du = \prod_\ell du_\ell \prod_m du_m du'_m$. Let

$$\eta_j^{(\ell)} = u_{j\ell}^{1/k}, \qquad \eta_j^{(m)} = u_{jm}^{1/2k}\exp\Big(\frac{i\psi_{jm}}{k}\Big), \quad 1\le j\le s-1.$$

The Jacobian of y_{ji} with respect to $u_{j\ell}, u_{jm}, \psi_{jm}$ is

$$|\det(w_j^{(i)})|^{-1}N\big(k^{-1}|\eta_j^{1-k}|\big) = D^{-\frac12}N\big(k^{-1}|\eta_j^{1-k}|\big).$$

We have

$$J(\Omega) = \int_Q \prod_\ell \chi_{\lambda_\ell}(u_\ell)\prod_m \chi_{\lambda_m}(u_m)\chi_{\lambda'_m}(u'_m)\,du\times$$

$$\times D^{-\frac12(1-s)}\int_R N\big(k^{-1}|\alpha_s^{-1}\eta_s^{1-k}|\big)\prod_{j=1}^{s-1}N\big(k^{-1}|\eta_j^{1-k}|\big)\prod_{j=1}^{s-1}\Big(\prod_\ell du_{j\ell}\prod_m du_{jm}d\psi_{jm}\Big),$$

where R denotes the region

$$o \le u_{ij} \le 1 \ (1 \le i \le s-1, \ 1 \le j \le r_1 + r_2),$$

$$-\pi \le \psi_{im} \le \pi \ (1 \le i \le s-1, \ r_1 + 1 \le m \le r_1 + r_2),$$

$$\alpha_s^{(\ell)} \eta_s^{(\ell)k} = z_\ell - \left(\alpha_1^{(\ell)} u_{1\ell} + \ldots + \alpha_{s-1}^{(\ell)} u_{s-1,\ell} - \mu^{(\ell)}\right), \ 0 \le \eta_s^{(\ell)} \le 1,$$

$$|\alpha_s^{(m)} \eta_s^{(m)k}| =$$
$$\left|z_m - \left(\alpha_1^{(m)} u_{1m}^{1/2} e^{i\psi_{1m}} + \cdots + \alpha_{s-1}^{(m)} u_{s-1,m}^{1/2} e^{i\psi_{s-1,m}} - \mu^{(m)}\right)\right|, \ |\eta_s^{(m)}| \le 1.$$

Therefore

$$J(\mu) = \lim_{\lambda_\ell, \lambda_m, \lambda'_m \to \infty} J(\Omega) = D^{\frac{1}{2}(1-s)} \prod_\ell F'_\ell \prod_m H'_m,$$

where

$$F'_\ell = \lim_{\lambda_\ell \to \infty} k^{-s} |\alpha_s^{(\ell)}|^{-1/k} \int_{Q_\ell} \chi_{\lambda_\ell}(u_\ell)\, du_\ell \times$$
$$\times \int_{W'_\ell} |u_\ell + \mu^{(\ell)} - \alpha_1^{(\ell)} w'_1 - \cdots - \alpha_{s-1}^{(\ell)} w'_{s-1}|^{\frac{1}{k}-1} \prod_{i=1}^{s-1} w_i'^{\frac{1}{k}-1} dw'$$

in which w'_i is used instead of $u'_{i\ell}$, $dw' = \prod_{1 \le i < s} dw'_i$, Q_ℓ denotes the region of u_ℓ in Q, and W'_ℓ the domain

$$0 \le w'_i \le 1, \ 1 \le i \le s, \quad \alpha_1^{(\ell)} w'_1 + \cdots + \alpha_s^{(\ell)} w'_s - \mu^{(\ell)} = u_\ell,$$

and where

$$H'_m = \lim_{\lambda_m, \lambda'_m \to \infty} k^{-2s} |\alpha_s^{(m)}|^{-2/k} \int_{Q_m} \chi_{\lambda_m}(u_m) \chi_{\lambda'_m}(u'_m) du_m\, du'_m \int_{V'_m} \times$$
$$\left|\frac{u_m - iu'_m}{\sqrt{2}} + \mu^{(m)} - \alpha_1^{(m)} w_1'^{\frac{1}{2}} e^{i\psi_1} - \cdots - \alpha_{s-1}^{(m)} w_{s-1}'^{\frac{1}{2}} e^{i\psi_{s-1}}\right|^{\frac{1}{k}-1} \prod_{i=1}^{s-1} w_i'^{\frac{1}{k}-1} dw'\, d\psi,$$

in which w'_i and ψ_i stand for u_{im} and ψ_{im} respectively, with $d\psi = \prod_{1 \le i < s} d\psi_i$, Q_m denotes the region of u_m and u'_m in Q, and V'_m the domain

$$0 \le w'_i \le 1, \ 1 \le i \le s, \qquad -\pi \le \psi_j \le \pi, \ 1 \le j \le s-1,$$

$$|\alpha_s^{(m)}|^2 w'_s = \left|\frac{u_m - iu'_m}{\sqrt{2}} + \mu^{(m)} - \alpha_1^{(m)} w_1'^{\frac{1}{2}} e^{i\psi_1} - \cdots - \alpha_{s-1}^{(m)} w_{s-1}'^{\frac{1}{2}} e^{i\psi_{s-1}}\right|^2.$$

By Lemma 6.8 and the transformation $|\alpha_i^{(\ell)}| w'_i = w_i \, (1 \le i \le s)$ for the integral F'_ℓ and $|\alpha_i^{(m)}|^2 w'_i = w_i \, (1 \le i \le s)$, $\phi_j = \theta_j + \psi_j$, where $\theta_j = \arg \alpha_i^{(m)}$ $(1 \le j \le s-1)$ for the integral H'_m we have

$$F'_\ell = k^{-s} \prod_{i=1}^{s} |\alpha_i^{(\ell)}|^{-1/k} F_\ell \quad \text{and} \quad H'_m = k^{-2s} \prod_{i=1}^{s} |\alpha_i^{(m)}|^{-2/k} H_m.$$

The theorem is proved. □

Notes

For Theorem 6.2, see Siegel [3,4], Tatuzawa [3] and Wang [1,2]. For Lemmas 6.6 and 6.8, see Tatuzawa [3].

Chapter 7

Waring's Problem

7.1 Introduction

Waring's problem in an algebraic number field is to consider the problem of decomposing a totally nonnegative integer ν as a sum of k-th powers of totally nonnegative integers, namely

$$\nu = \lambda_1^k + \cdots + \lambda_s^k, \tag{7.1}$$

where $(\lambda_1, \ldots, \lambda_s) \in P^s$. It was pointed out by Siegel that there may exist infinitely many integers in P which are not sums of k-th powers. This led him to consider the ring J_k generated by k-th powers of integers instead of P; see Introduction. Suppose that $\nu \in J_k \cap P$. Let $r_s(\nu)$ be the number of solutions of the equation (7.1) subject to the condition

$$\lambda_i \in P(T), \qquad 1 \leq i \leq s.$$

Then $r_s(\nu)$ can be expressed as an integral over U_n as in §5.1. The corresponding singular series is

$$\mathfrak{S}(\nu) = \sum_{\gamma} G(\gamma) E(-\nu\gamma),$$

where

$$G(\gamma) = \Big(N(\mathfrak{a}) \sum_{\lambda(\mathfrak{a})} E(\lambda^k \gamma)\Big)^s, \quad \gamma \to \mathfrak{a},$$

and we shall see that the singular integral is given by

$$J(\mu) = D^{(1-s)/2} k^{-sn} \Gamma\Big(\frac{1}{k}\Big)^{sr_1} \Gamma\Big(\frac{s}{k}\Big)^{-r_1} \prod_{\ell} \mu^{(\ell)s/k-1} \prod_{m} H_m,$$

where $\mu = \nu T^{-k}$ and

$$H_m = \int_{V_m} \prod_{i=1}^{s} w_i^{1/k-1} \, dw \, d\varphi,$$

with $dw = \prod_{1 \leq i < s} dw_i$, $d\varphi = \prod_{1 \leq i < s} d\varphi_i$ and V_m being the region determined by

$$0 \le w_i \le 1, \quad 1 \le i \le s, \qquad -\pi \le \varphi_j \le \pi, \quad 1 \le j \le s-1,$$
$$w_s = \left|\mu^{(m)} - w_1^{1/2} e^{i\varphi_1} - \cdots - w_{s-1}^{1/2} e^{i\varphi_{s-1}}\right|^2.$$

Let $G_{\mathbb{K}}(k)$ be the least value of s for which every ν in $J_k \cap P$, with $N(\nu)$ sufficiently large, is representable in the form (7.1) with $(\lambda_1, \ldots, \lambda_s) \in P^s$ subject to the condition

$$N(\lambda_i)^k \ll N(\nu), \qquad 1 \le i \le s.$$

If $\mathbb{K} = \mathbb{Q}$, then $G_{\mathbb{K}}(k)$ is simply $G(k)$.

Let sP^k denote the set of integers of the form

$$\nu = \lambda_1^k + \cdots + \lambda_s^k,$$

where $(\lambda_1, \ldots, \lambda_s) \in P^s$. For any finite set $F \subset P$, let $N(\lambda : \lambda \in F)$ be the number of elements of F.

Theorem 7.1. *If $s \ge 4kn$ and $\nu \in J_k$, then there exists a positive constant $c_1(s, \mathbb{K})$ such that $\mathfrak{S}(\nu) > c_1(s, \mathbb{K})$.*

Theorem 7.2. *If $s \ge \max(4kn, 2^k + 1)$ and $\mu = \nu T^{-k} \in P(1)$, then*

$$r_s(\nu) = \mathfrak{S}(\nu) J(\mu) T^{(s-k)n} + O\left(T^{(s-k)n-2^{-k}}\right).$$

Theorem 7.3. *We have*

$$G_{\mathbb{K}}(k) \le 2nG(k) + 4kn.$$

Theorem 7.4. *We have*

$$\frac{N\left(\lambda : \lambda \in P(x) \cap \frac{1}{2} 8^{k-1} P^k\right)}{N(\lambda : \lambda \in P(x) \cap J_k)} \ge c_2(k, \mathbb{K}) > 0.$$

Theorem 7.4 tells us that the set $\frac{1}{2} 8^{k-1} P^k$ has a positive density in $P \cap J_k$. In paricular, the set of sums of four squares of numbers in P has a positive density in $P \cap J_2$.

7.2 The Ring J_k

Lemma 7.1. *Let a_{ij}, $1 \le i \le r$, $1 \le j \le s$ be rational integers. If $s > r$, then the system of congruences*

$$\sum_{j=1}^{s} a_{ij}x_j \equiv 0 \pmod{p^t}, \qquad 1 \le i \le r$$

has a nontrivial solution in rational integers $x_1, \ldots, x_s$, in the sense that at least one x_i is not a multiple of p.

Proof. Suppose that $(a_{1j}, \ldots, a_{rj}) = d_j$, $1 \le j \le s$ and

$$p^m \| (d_1, \ldots, d_s)$$

for $0 \le m < t$, since the lemma is obvious for the case $m \ge t$. Let $a_{ij} = p^m a'_{ij}$. Then the congruences become

$$\sum_{j=1}^{s} a'_{ij}x_j \equiv 0 \pmod{p^{t-m}}, \qquad 1 \le i \le r.$$

We may assume without loss of generality that $p \not| a'_{11}$. Then the congruences can be reduced to

$$\sum_{j=1}^{s} a'_{1j}x_j \equiv 0 \pmod{p^{t-m}},$$

$$\sum_{j=2}^{s} b_{ij}x_j \equiv 0 \pmod{p^{t-m}}, \qquad 2 \le i \le r,$$

and the lemma is easily proved by induction on r. □

Lemma 7.2. *Let ρ be a primitive root* $\bmod \wp$. *Then any α of J is expressible as*

$$\alpha \equiv \sum_{i=0}^{e-1} \sum_{j=0}^{f-1} a_{ij}\rho^j \pi^i \pmod{\wp^{te}}$$

for any positive integer t, and the a_{ij} are uniquely determined $\bmod p$, *and $\wp \| \pi$.*

Proof. Since a complete residue system $\bmod \wp$ forms a finite field, it follows that for any $\beta \in J$, there exist rational integers c_{0j}, unique $\bmod \wp$, such that

$$\beta \equiv \sum_{j=0}^{f-1} c_{0j}\rho^j \pmod{\wp}.$$

Furthermore, for any $\alpha \in J$ we can determine $c_{rj} \pmod{p}$ such that

$$\alpha \equiv \sum_{r=0}^{te-1} \sum_{j=0}^{f-1} c_{rj} \rho^j \pi_r \pmod{\wp^{te}},$$

where π_j are integers satisfying $\wp^j \| \pi_j$. Let $r = qe + i$, $0 \le i < e$. If we put

$$\pi_r = p^q \pi^i,$$

then the above congruence becomes

$$\begin{aligned} \alpha &\equiv \sum_{q=0}^{t-1} p^q \sum_{i=0}^{e-1} \sum_{j=0}^{f-1} c_{qe+i,j} \rho^j \pi^i \\ &\equiv \sum_{i=0}^{e-1} \sum_{j=0}^{f-1} a_{ij} \rho^j \pi^i \qquad \pmod{\wp^{te}}, \end{aligned}$$

where

$$a_{ij} = \sum_{q=0}^{t-1} c_{qe+i,j} p^q, \qquad 0 \le i \le e-1, \ \ 0 \le j \le f-1$$

are uniquely determined $\bmod p$. The lemma is proved. □

Lemma 7.3. *We have*

$$k!\, x = \sum_{i=0}^{k-1} (-1)^{k-1-i} \binom{k-1}{i} \left((x+i)^k - i^k \right).$$

Proof. We first prove that

$$\sum_{i=0}^{s} (-1)^i i^j \binom{s}{i} = \begin{cases} 0, & \text{if } \ 0 \le j < s, \\ (-1)^s s!, & \text{if } \ j = s. \end{cases}$$

In fact, by differentiating the expression

$$(1+x)^s = \sum_{i=0}^{s} \binom{s}{i} x^i$$

several times, we obtain

$$\frac{s!}{(s-j)!} (1+x)^{s-j} x^j + (1+x)^{s-j+1} f(x) = \sum_{i=0}^{s} i^j \binom{s}{i} x^i,$$

where $f(x)$ is a polynomial with integral coefficients of degree $j-1$. The assertion follows by putting $x = -1$. Hence

$$\begin{aligned}
&\sum_{i=0}^{k-1}(-1)^{k-1-i}\binom{k-1}{i}((x+i)^k-i^k)\\
=&\sum_{i=0}^{k-1}(-1)^{k-1-i}\binom{k-1}{i}\sum_{j=1}^{k}\binom{k}{j}x^j i^{k-j}\\
=&(-1)^{k-1}\sum_{j=1}^{k}\binom{k}{j}x^j\sum_{i=0}^{k-1}(-1)^i\binom{k-1}{i}i^{k-j}\\
=&(-1)^{k-1}\binom{k}{1}x(-1)^{k-1}(k-1)!=k!\,x.
\end{aligned}$$

The lemma is proved. □

It follows by Lemma 7.3 that the ring J_k contains the principal ideal $(k!)$. Since also 1 belongs to J_k, the ring J_k is an order.

Lemma 7.4 (Siegel). *Let t be a positive rational integer. Then there exists rational integers $s\,(\le tf)$, q_j (some powers of p) and integers $\eta_i\,(1\le i\le s)$ such that the linear forms*

$$a_1\eta_1^k+\cdots+a_s\eta_s^k,\qquad 0\le a_j\le q_j-1,\quad 1\le j\le s$$

uniquely represent all numbers of J_k $\bmod \wp^t$.

Proof. Let q denote the order of the module generated by the residue classes represented by elements of J_k $\bmod \wp^t$. Take $\eta_1\notin\wp$. The set

$$\{x:\ x \text{ rational integer},\ x\eta_1^k\equiv 0\ (\bmod\ \wp^t)\}$$

becomes a module and contains p^t. Hence the smallest positive integer in this set, say q_1, is a power of p. Next, we take η_2 such that

$$\eta_2\not\equiv a_1\eta_1^k\ (\bmod\ \wp^t)\qquad\text{for}\quad a_1=0,1,\ldots,q-1.$$

If no such η_2 exists, then $q=q_1$. Next we consider the set

$$\{\,x\ :\ x\eta_2^k\equiv a_1\eta_1^k\ (\bmod\ \wp^t)\quad\text{for some } a_1\,\}.$$

This is also a module and contains p^t. Hence the smallest positive integer in this set, say q_2, is a power of p. Then the numbers

$$a_1\eta_1^k+a_2\eta_2^k,\qquad 0\le a_i\le q_i-1,\quad i=1,2$$

are relatively incongruent $\bmod\ \wp^t$. Then we take η_3 such that

$$\eta_3^k\not\equiv a_1\eta_1^k+a_2\eta_2^k\ (\bmod\ \wp^t),\qquad 0\le a_i\le q_i-1,\ i=1,2.$$

If no such η_3 exists, then $q=q_1q_2$.

In this way we can infer that $q = q_1 \cdots q_s$, and any number in J_k can be uniquely expressed as

$$a_1\eta_1^k + \cdots + a_s\eta_s^k \pmod{\wp^t}, \qquad 0 \le a_j \le q_j - 1,\ 1 \le j \le s.$$

Since $p^s \le q_1 \cdots q_s = q \le N(\wp)^t = p^{tf}$, the lemma follows. □

Lemma 7.5. *Under the same assumption as in* Lemma 7.4, *we have $s \le ef$, for whatever t.*

Proof. For simplicity, we denote the basis $\pi^i\rho^j \pmod{\wp^{te}}$ in Lemma 7.2 by

$$\sigma_1, \cdots, \sigma_r \quad (r = ef).$$

As to η_j in Lemma 7.4 we can determine rational integers a_{ij} such that

$$\eta_j^k \equiv \sum_{i=1}^{r} a_{ij}\sigma_i \pmod{\wp^{te}}, \qquad 1 \le j \le s.$$

If $s > r$, then it follows by Lemma 7.1 that there exist rational integers x_i, not all divisible by p, such that

$$\sum_{j=1}^{s} a_{ij}x_j \equiv 0 \pmod{p^t}, \qquad 1 \le i \le r.$$

This leads to the contradictary

$$\sum_{i=1}^{s} x_i\eta_i^k \equiv 0 \pmod{\wp^{te}}$$

and therefore $s \le r \le ef$. The lemma is proved. □

7.3 Proofs of Theorems 7.1 and 7.2

Proof of Theorem 7.1. Let $t_1 = \gamma e$ and $s_0 = 4kef\,(\le 4kn)$, where γ is defined in §1.6. Then, by Lemmas 1.17 and 7.5, it follows that for any $\nu \in J_k$ the congruence

$$\lambda_1^k + \cdots + \lambda_s^k \equiv \nu \pmod{\wp^{t_1}}$$

is solvable with $\wp \nmid \lambda_1$ provided $s \ge s_0$. The theorem follows by Lemmas 6.4 and 6.7 since $t \ge t_0$. □

Proof of Theorem 7.2. The theorem follows immediately from Lemma 1.10 and Theorems 5.2 and 6.2. □

7.4 Proof of Theorem 7.3

Lemma 7.6. *Let T_1 be a sufficiently large positive number and $s_1 = nG(k)$. Let $W_{s_1}(T_1)$ denote the set of integers in $\mathbb{K}$ which can be expressed in the form*

$$\lambda_1^k + \cdots + \lambda_{s_1}^k,$$

where λ_i are integers satisfying $\lambda_i \in P(T_1)$, $1 \le i \le s$, and let $N_{s_1}(T_1)$ be the number of integers belonging to $W_{s_1}(T_1)$. Then we have

$$N_{s_1}(N_1) \ge c_3(k, \mathbb{K}) T_1^{kn}.$$

Proof. Take an integer θ such that

$$1, \theta, \ldots, \theta^{n-1}$$

are linearly independent over $\mathbb{Q}$. Let ζ_k denote a k-th root of unity. Then we can take a large rational integer z such that $\theta + z \in P$ and

$$\theta^{(i)} + z \ne (\theta^{(j)} + z)\zeta_k, \qquad 1 \le i < j \le n$$

for every ζ_k. It follows that

$$1, (\theta + z)^k, \ldots, (\theta + z)^{k(n-1)}$$

are totally nonnegative and linearly independent over $\mathbb{Q}$, on account of

$$\det \begin{vmatrix} 1 & (\theta^{(1)} + z)^k & \ldots & (\theta^{(1)} + z)^{k(n-1)} \\ \vdots & \vdots & & \vdots \\ 1 & (\theta^{(n)} + z)^k & \ldots & (\theta^{(n)} + z)^{k(n-1)} \end{vmatrix} = \prod_{1 \le i < j \le n} \left((\theta^{(i)} + z)^k - (\theta^{(j)} + z)^k\right) \ne 0.$$

Let x_i, $0 \le i \le n-1$ be rational integers satisfying

$$x_i(\theta + z)^{ik} \in P(T_1^k), \qquad 0 \le i \le n-1.$$

It follows from the definition of $G(k)$ that almost all numbers of the form

$$x_0 + x_1(\theta_1 + z)^k + \cdots + x_{n-1}(\theta + z)^{k(n-1)}$$

belong to $W_{s_1}(T_1)$, and the lemma follows. □

Proof of Theorem 7.3. Take $s = 4kn$. For a given nonzero number ν of $J_k \cap P$, we can take in Lemma 3.1 $\gamma_i = k^{-1}\left(\log N(\nu)^{1/n} + \log |\nu^{(i)}|^{-1}\right)$, $1 \le i \le s$ and a unit $\eta \in P$ such that $e^{\gamma_i} \ll |\eta^{(i)}| \ll e^{\gamma_i}$, $1 \le i \le n$. Then $\chi = \nu\eta^k$ satisfies

$$c_4 M \le \chi^{(\ell)} \le c_5 M, \qquad c_4 M \le |\chi^{(m)}| \le c_5 M,$$

where $c_4 = c_4(k, \mathbb{K})\, c_5 = c_5(k, \mathbb{K})$ and $M = \sqrt[n]{N(\nu)}$. Take σ_1, σ_2 in $W_{s_1}(T_1)$ with

$$T_1 = \left(\frac{c_4 M}{4s_1}\right)^{1/k}$$

and set $\tau = \chi - \sigma_1 - \sigma_2$ and $\mu = T^{-k}\tau$, where

$$T = \left((c_5 + \tfrac{1}{2}c_4)M\right)^{1/k}.$$

Then

$$c_6 = (c_5 + \tfrac{c_4}{2})^{-1}(c_4 - \tfrac{c_4}{2}) \le \mu^{(\ell)} \le (c_5 + \tfrac{c_4}{2})^{-1}(c_4 + \tfrac{c_4}{2}) = 1 \text{ and } c_6 \le |\mu^{(m)}| \le 1.$$

Therefore $J(\mu) > c_7(k, \mathbb{K}) > 0$. Let

$$L(\xi) = \sum_{\lambda \in P(T)} E(\lambda^k \xi) \quad \text{and} \quad V(\xi) = \sum_{\sigma \in W_{s_1}(T_1)} E(\sigma\xi).$$

It follows from Theorem 5.1 that

$$\begin{aligned}\int_B L(\xi)^{4kn} V(\xi)^2 E(-\chi\xi)\, dx &= \sum_\tau \int_B L(\xi)^{4kn} E(-\tau\xi)\, dx \\ &= \sum_\tau \mathfrak{S} J(\mu) T^{(4kn-k)n}\big(1 + o(1)\big).\end{aligned}$$

This yields

$$\operatorname{Re} \int_B L(\xi)^{4kn} V(\xi)^2 E(-\chi\xi)\, dx > c_8(k, \mathbb{K}) T^{(4kn-k)n} N_{s_1}(T_1)^2.$$

On the other hand, by Lemma 5.2 and Lemma 5.10 with $\theta = 2^{-k}$, we have

$$\begin{aligned}\int_S L(\xi)^{4kn} V(\xi)^2 E(-\chi\xi)\, dx &\ll T^{4kn(n-2^{-k})} \int_S |V(\xi)|^2 dx \\ &\ll T^{4kn(n-2^{-k})} \int_{U_n} \sum_{\sigma_1, \sigma_2 \in W_{s_1}(T_1)} E\big((\sigma_1 - \sigma_2)\xi\big)\, dx \\ &\ll T^{4kn(n-2^{-k})} N_{s_1}(T_1).\end{aligned}$$

Therefore

$$\int_{U_n} L(\xi)^s V(\xi)^2 E(-\chi\xi)\, dx > 0$$

if M is large. This means that $\chi = \nu\eta^k$ is expressible as

$$\nu\eta^k = \lambda_1^k + \cdots + \lambda_{4kn}^k + \tau_1^k + \cdots + \tau_{2s_1}^k,$$

subject to the conditions

$$\lambda_i \in P(T),\ 1 \le i \le 4kn \quad \text{and} \quad \tau_j \in P(T_1),\ 1 \le j \le 2s_1.$$

The theorem is proved. □

7.5 Proof of Theorem 7.4

A polynomial $f(\lambda)$ is said to be positive if $f(\lambda) \in P$ for all $\lambda \in P$. For any polynomial $f(\lambda)$ with coefficients in J and any sets $F_i \subset P$, $1 \le i \le s$, we define the set

$$\sum_{i=1}^{s} f(F_i) = \{\, \nu : \nu = f(\lambda_1) + \cdots + f(\lambda_s),\ \lambda_i \in F_i,\ 1 \le i \le s \,\}. \tag{7.2}$$

When $f(\lambda) = \lambda^k$ and $F_i = F$, $1 \le i \le s$, the left hand side is sF^k; see §7.1. Write

$$F[x] = \sum_{i=1}^{s} f\big(F_i \cap P(x)\big) \quad \text{and} \quad P[x] = \sum_{i=1}^{s} f\big(P(x)\big).$$

Let $h(\nu, x, F)$ and $h(\nu, x, P)$ denote the respective numbers of solutions of the equation

$$\nu = f(\lambda_1) + \cdots + f(\lambda_s)$$

satisfying

$$\lambda_i \in F_i \cap P(x) \quad \text{and} \quad \lambda_i \in P(x),\ 1 \le i \le s.$$

For any set $F \subset P$, the Schnirelman density of F is defined by

$$d(F) = \inf_{x \ge 1} \frac{N(\lambda : \lambda \in P(x) \cap F)}{N(\lambda : \lambda \in P(x))}.$$

We have $0 \le d(F) \le 1$, and by Lemma 3.2

$$N(\lambda : \lambda \in P(x) \cap F) \ge c_9(\mathbb{K}) x^n d(F), \quad x \ge 1. \tag{7.3}$$

Lemma 7.7. *Let $s = \frac{1}{2}8^{k-1}$ and*

$$f(\lambda) = \alpha_k \lambda^k + \cdots + \alpha_1 \lambda$$

be a polynomial of degree k with coefficients in J. Suppose that $F_i \subset P$ and $0 \in F_i$, $1 \le i \le s$. Then

$$d\Big(\sum_{i=1}^{s} f(F_i)\Big) \ge c_{10}(k, \mathbb{K}) \prod_{i=1}^{s} d(F_i)^2.$$

Proof. By (7.3) we have

$$\sum_{\nu \in F[x]} h(\nu, x, F) = \prod_{i=1}^{s} N(\lambda : \lambda \in F_i \cap P(x)) \ge c_9^s x^{sn} \prod_{i=1}^{s} d(F_i). \tag{7.4}$$

Since

$$F[x] \subset P\big(c_{11}(k, \mathbb{K}) x^k\big) \cap \sum_{i-1}^{s} f(F_i),$$

it follows by Schwarz's inequality that

$$\Big(\sum_{\nu\in F[x]} h(\nu,x,F)\Big)^2 \le N(\lambda:\lambda\in F[x]) \sum_{\nu\in F[x]} h(\nu,x,F)^2$$

$$\le N\big(\lambda:\lambda\in P(c_{11}x^k)\cap\sum_{i=1}^{s} f(F_i)\big) \sum_{\nu\in P[x]} h(\nu,x,P)^2.$$

By Theorem 4.2,

$$\sum_{\nu\in P[x]} h(\nu,x,P)^2 = \int_{U_n}\Big|\sum_{\lambda\in P(x)} E\big(f(\lambda)\xi\big)\Big|^{2s} dx \ll x^{(2s-k)n},$$

and therefore by (7.4),

$$c_9^{2s}\prod_{i=1}^{s} d(F_i)^2 \le N\big(\lambda:\lambda\in P(c_{11}x^k)\cap\sum_{i=1}^{s} f(F_i)\big)c_{12}x^{-kn}, \qquad x\ge 1.$$

Set $y = c_{11}x^k$. We then have

$$N\big(\lambda:\lambda\in P(y)\cap\sum_{i=1}^{s} f(F_i)\big) \ge c_{13}y^n\prod_{i=1}^{s} d(F_i)^2, \quad y\ge c_{11}. \tag{7.5}$$

Since $0\in P(y)\cap\sum_{i=1}^{s} f(F_i)$, we have

$$N\big(\lambda:\lambda\in P(y)\cap\sum_{i=1}^{s} f(F_i)\big) \ge 1 \ge \frac{y^n}{c_{11}^n}\prod_{i=1}^{s} d(F_i)^2, \quad y\le c_{11}. \tag{7.6}$$

The lemma follows by (7.5) and (7.6). □

Set $F_i = P$, $1\le i\le s$ and $f(\lambda)=\lambda^k$. Then $d(F_i)=1$, $1\le i\le s$, and Theorem 7.4 follows by Lemma 7.7. □

Notes

Siegel [3,4] first proved that $\mathfrak{S}(\nu)>0$ if $s>(2^{k-1}+n)kn$. Siegel's result was improved to $s\ge[8k(\log k+1)n]$ by Tatuzawa [1]. Theorem 7.1 was established by Stemmler [1] and Tatuzawa [3].

Siegel [3,4] first established the asymptotic formula for $r_s(\nu)$ if $s>(2^{k-1}+n)kn$, and it derives immediately that $G_{\mathbf{K}}(k)\le(2^{k-1}+n)kn+1$. Using Vinogradov's method in algebraic number fields, the estimations for $G_{\mathbf{K}}(k)$ were improved successively by Tatuzawa, Körner and Eda. Their upper bounds (Tatuzawa [1], Körner [2] and Eda [3]) for $G_{\mathbf{K}}(k)$ are as follows:

$$8k(k+n)n,$$

$$k\left(3\log k + 3\log\left(\tfrac{n^2+1}{3}\right) + 1\right)n,$$

$$2kn\log k + 6kn\log\log k + 2k(2n\log n + n\log\log n + 2\log n + 14) + 1;$$

the last bound being valid for $k \geq c_{14}(\mathbb{K})$. Theorem 7.3 was proved by Tatuzawa [3], where the best record for $G_{\mathbb{K}}(k)$, namely $G_{\mathbb{K}}(k) \sim 2k\log k$, is due to Vinogradov; see [Vinogradov 3] and [Vinogradov 1].

Theorem 7.4: See Rieger [1,2].

Lemma 7.4: See Siegel [3].

Very recently T. D. Wooley [1] has made a substantial improvement on the upper estimate of $G(k)$. In a paper submitted to Annals of Mathematics he proved that $G(k) \leq k\big(\log k + \log\log k + O(1)\big)$, and there are also extraneous results for smaller values of k.

Chapter 8

Additive Equations

8.1 Introduction

Let $\alpha_i\,(1 \leq i \leq s)$ be a set of nonzero integers. The form

$$A(\boldsymbol{\lambda}) = \sum_{i=1}^{s} \alpha_i \lambda_i^k$$

is called an *additive form*, and the equation

$$A(\boldsymbol{\lambda}) = 0 \tag{8.1}$$

its corresponding *additive equation*. Let $R_s(0)$ be the number of solutions of (8.1) subject to the condition: $\lambda_i \in P(T), 1 \leq i \leq s$. Then $R_s(0)$ can be expressed as an integral over U_n; see §5.1. The corresponding singular series is

$$\mathfrak{S}(0) = \sum_{\gamma} G(\gamma) = \sum_{\mathfrak{a}} H(\mathfrak{a}),$$

where

$$G(\gamma) = \prod_{i=1}^{s} G_i(\gamma),$$

$$G_i(\gamma) = N(\mathfrak{a}_i)^{-1} \sum_{\lambda(\mathfrak{a}_i)} E(\alpha_i \lambda^k \gamma), \qquad \gamma\alpha_i\delta \to \mathfrak{a}_i, \quad 1 \leq i \leq s,$$

and where

$$H(\mathfrak{a}) = \sum_{\gamma}{}^{*} G(\gamma)$$

in which γ runs over a reduced system of $(\mathfrak{a}\delta)^{-1} \bmod \delta^{-1}$.

Theorem 8.1. *If* $s \geq (2k)^{n+1}$, *then*

$$\mathfrak{S}(0) > c_1(s, A, \mathbb{K}) > 0,$$

where $A = \max_i \|\alpha_i\|$.

Theorem 8.2. *If k is odd and $s \geq ckn \log k$, then*

$$\mathfrak{S}(0) > c_2(s, A, \mathbb{K}) > 0.$$

Theorem 8.3. *Suppose that $s \geq c_3(k,n)$ and that, for each ℓ, the numbers $\alpha_i^{(\ell)}$ $(1 \leq i \leq s)$ do not have the same sign. Then the equation* (8.1) *has a solution $\boldsymbol{\lambda} \in P^s$ such that $\boldsymbol{\lambda} \neq \mathbf{0}$ and*

$$\|\boldsymbol{\lambda}\| \ll A^{c_4(k,n)}. \tag{8.2}$$

The following two theorems can then be derived from Theorem 8.3.

Theorem 8.4. *If $s \geq c_3(k,n)$ and $r_1 = 0$, so that $\mathbb{K}$ is a purely imaginary algebraic number field, then the conclusion of* Theorem 8.3 *holds.*

Theorem 8.5. *If k is odd and $s \geq c_3(k,n)$, then the equation* (8.1) *has a solution $\boldsymbol{\lambda} \neq \mathbf{0}$ in J^s satisfying* (8.2).

8.2 Reductions

Let $R(\mathfrak{a})$ be a given set of complete residue system mod $\mathfrak{a}$. Any integer α of $\mathbb{K}$ has a unique $\wp$-adic representation

$$\alpha = \sum_{i=0}^{\infty} \mu_i \pi^i, \qquad \mu_i \in R(\wp),$$

where $\wp \| \pi$. In fact, μ_0 is uniquely determined by $\mu_0 \equiv \alpha \pmod{\wp}$, and then μ_1 is determined by $\pi\mu_1 \equiv \alpha - \mu_0 \pmod{\wp^2}$, and so on.

Let $M(\mathfrak{a})$ be the number of solutions of the congruence

$$A(\boldsymbol{\lambda}) \equiv 0 \pmod{\mathfrak{a}}$$

subject to $\lambda_i \in R(\mathfrak{a})$, $1 \leq i \leq s$. We write

$$\alpha_i = \beta_i \pi^{h_i k + t_i}, \qquad \wp \nmid \beta_i, \quad 0 \leq t_i < k, \quad 1 \leq i \leq s.$$

Then the congruence

$$A(\boldsymbol{\lambda}) \equiv 0 \pmod{\wp^v}$$

becomes

$$\sum_{i=1}^{s} \beta_i \pi^{t_i} (\pi^{h_i} \lambda_i)^k \equiv 0 \pmod{\wp^v}.$$

Let $h = \max_i h_i$. We restrict ourselves to solutions of the form

$$\lambda_i = \pi^{h-h_i}\mu_i, \qquad 1 \le i \le s.$$

Then, for large v, we can cancel π^{hk} from the congruence, which then becomes

$$\sum_{i=1}^{s} \beta_i \pi^{t_i} \mu_i^k \equiv 0 \pmod{\wp^{v-hk}}, \tag{8.3}$$

where $\mu_i \in R(\wp^{v-h+h_i})$, $1 \le i \le s$. If we denote by $M_1(\wp^{v-hk})$ the number of solutions of (8.3) subject to

$$\mu_i \in R(\wp^{v-hk}), \qquad 1 \le i \le s,$$

then we obtain

$$M(\wp^v) \ge M_1(\wp^{v-hk}).$$

We group together the terms according to the values t_i in (8.3). There are k groups, and one at least of these must contain $u \ge s/k$ terms. We assume that the group with $t_i = t$ has at least s/k terms. Then (8.3) becomes

$$B^{(0)} + \pi B^{(1)} + \cdots + \pi^{k-1} B^{(k-1)} \equiv 0 \pmod{\wp^{v-hk}},$$

where $B^{(i)}$ is an additive form with all coefficients not congruence to 0 mod $\wp$, the variables in distinct forms $B^{(i)}$ are distinct and the number of variables in $B^{(t)}$ is $u \ge s/k$. Set $\mu_i = \pi\nu_i$ for variables in $B^{(0)}, \ldots, B^{(t-1)}$ and cancel the factor π^t. The above congruence becomes

$$F = F^{(0)} + \pi F^{(1)} + \cdots + \pi^{k-1} F^{(k-1)} \equiv 0 \pmod{\wp^{v-hk-t}}, \tag{8.4}$$

where $F^{(i)}$ is an additive form in v_i variables (the variables in distinct forms $F^{(i)}$ being distinct) with all coefficients not congruent to 0 mod $\wp$, and where $v_0 \ge s/k$.

Let $M_2(\mathfrak{a})$ be the number of solutions of

$$F(\boldsymbol{\lambda}) \equiv 0 \pmod{\mathfrak{a}}$$

subject to $\lambda_i \in R(\mathfrak{a})$, $1 \le i \le s$. Then we have, for $v > (h+1)k$,

$$M(\wp^v) \ge M_2(\wp^{v-(h+1)k}). \tag{8.5}$$

8.3 Contraction

Lemma 8.1. *Let $d = (k, N(\wp) - 1)$. Suppose that $\wp \nmid \alpha_i$, $1 \le i \le d+1$. Then*

$$\alpha_1 \lambda_1^k + \cdots + \alpha_{d+1} \lambda_{d+1}^k \equiv 0 \pmod{\wp}$$

has a solution with $\wp \nmid \lambda_1$.

This lemma is a special case of Chevalley's theorem on diophantine equations over finite fields.

Proof. It is known that the congruences

$$\lambda^k \equiv \mu \pmod{\wp} \qquad \lambda^d \equiv \mu \pmod{\wp}$$

are either both soluble or both not soluble; see §2.6. Thus, if the conclusion of the lemma does not hold, then, for any $\lambda_2, \ldots, \lambda_{d+1}$, we have

$$\alpha_1 + \alpha_2\lambda_2^d + \cdots + \alpha_{d+1}\lambda_{d+1}^d \not\equiv 0 \pmod{\wp},$$

that is,

$$\left(\alpha_1 + \alpha_2\lambda_2^d + \cdots + \alpha_{d+1}\lambda_{d+1}^d\right)^{N(\wp)-1} \equiv 1 \pmod{\wp}.$$

This should be an identity. But the expansion of the left hand side contains a term

$$\alpha\left(\lambda_2^d \cdots \lambda_{d+1}^d\right)^{\frac{N(\wp)-1}{d}} = \alpha(\lambda_2 \cdots \lambda_{d+1})^{N(\wp)-1},$$

where $\wp \nmid \alpha$, and every other term contains at least one of the variables $\lambda_2, \ldots, \lambda_{d+1}$ to a power less than $N(\wp) - 1$. This leads to a contradiction, and the lemma is proved. □

We now define an operation of *contraction* for a form of the type (8.4) as follows: Consider a sum of $d+1$ terms

$$\alpha_1\lambda_1^k + \cdots + \alpha_{d+1}\lambda_{d+1}^k \tag{8.6}$$

in $F^{(0)}$. Since $\wp \nmid \alpha_i$, $1 \le i \le d+1$, the congruence

$$\alpha_1\mu_1^k + \cdots + \alpha_{d+1}\mu_{d+1}^k \equiv 0 \pmod{\wp} \tag{8.7}$$

has a solution with $\wp \nmid \mu_1$ by Lemma 8.1. By choosing the solution suitably we can assume that the integer on the left hand side of (8.7) is a nonzero element of α. Set $\lambda_i = \mu_i\lambda$, $1 \le i \le d+1$. Then

$$\alpha_1\lambda_1^k + \cdots + \alpha_{d+1}\lambda_{d+1}^k = \left(\alpha_1\mu_1^k + \cdots + \alpha_{d+1}\mu_{d+1}^k\right)\lambda^k = \alpha\lambda^k, \tag{8.8}$$

where $\wp^j \| \alpha$, $j \ge 1$. Thus α has a $\wp$-adic representation

$$\alpha = \pi^j\nu_j + \cdots, \qquad \wp \nmid \nu_j.$$

This is the operation of contraction. The sum (8.6) is contracted into a single term (8.8). Note that $\lambda \not\equiv 0 \pmod{\wp}$ implies that $\lambda_1 \not\equiv 0 \pmod{\wp}$, where λ_1 is called the *distinguished variable* in (8.6).

Operations of contraction are applied to groups of $d+1$ terms in $F^{(0)}$, and here any one of the variables can be chosen as the distinguished variable. The number of remaining variables is then at most d, and we put them equal to zero. Therefore there is a form of the type

$$G = \pi G^{(1)} + \pi^2 G^{(2)} + \cdots,$$

where $G^{(j)}$ contains the original v_j terms of $F^{(j)}$ together with possibly some additional terms, each arising from the form (8.8) from one of the contractions. The variables in these additional terms are called the *derived variables*. Here we put $v_j = 0$ if $j \geq k$.

The next step will be to apply contractions to groups of $d+1$ terms in $G^{(1)}$, subject to the condition that each group contains at least one term with a derived variable; such a variable is chosen as the distinguished variable in contraction. The process is continued, and at each stage we take care that at least one of the varibles in a group is derived, either directly or indirectly from a variable in $F^{(0)}$. Suppose that, after the u-th contraction, we obtain

$$H = \pi^u H^{(u)} + \pi^{u+1} H^{(u+1)} + \cdots, \tag{8,9}$$

where $H^{(u)}, H^{(u+1)}, \ldots$ are additive forms in distinct variables with all coefficients not congruent to 0 $(\bmod \wp)$. For $j \leq k-1$, the form $H^{(j)}$ includes the terms of $F^{(j)}$, and possibly some additional terms containing derived variables; for $j \geq k$, the form $H^{(j)}$ can contain only derived variables.

8.4 Derived Variables

Lemma 8.2. *Let S_u be the total number of derived variables in the form* (8.9). *Then*

$$S_u \geq \frac{v_0}{(d+1)^u} - 1.$$

Proof. Suppose first that $u = 1$. We divide the v_0 terms of $F^{(0)}$ into h sets of $d+1$ terms, where

$$h = \left[\frac{v_0}{d+1}\right] \geq \frac{v_0}{d+1} - \frac{d}{d+1} > \frac{v_0}{d+1} - 1,$$

and the lemma is true. Suppose now that $u \geq 1$, and that the lemma holds for u. We proceed to show that the lemma holds for $u+1$. Let w denote the number of derived variables in $H^{(u)}$. Then the total number of derived variables in $H^{(u+1)}, \ldots$ is $S_u - w$. We divide the $v_u + w$ terms in $H^{(u)}$ into as many sets of $d+1$ terms as possible subject to the condition that each set contains at least one of the derived variables. The number of sets that can be formed is

$$\begin{cases} w & \text{if} \quad v_u \geq dw, \\ \left[\frac{v_u + w}{d+1}\right] & \text{if} \quad v_u < dw. \end{cases}$$

Any variables remaining in $H^{(u)}$ are equated to zero. Each set of $d+1$ terms can be contracted into a single term

$$\pi^g \alpha \lambda^k, \qquad g \geq u+1, \quad \wp \not| \alpha.$$

Adding all such terms to the corresponding form $\pi^g H^{(g)}$ we obtain a form of the type

$$\pi^{u+1} P^{(u+1)} + \pi^{u+2} P^{(u+2)} + \cdots,$$

where each $P^{(j)}$ contains the variables in $H^{(j)}$ and new derived variables of the form $\pi^j \alpha \lambda^k$, $\wp \not| \alpha$. The total number of derived variables in $P^{(u+1)}, \ldots$ is

$$S_u - w + \begin{cases} w & \text{if} \quad v_u \geq dw, \\ \left[\frac{v_u + w}{d+1}\right] & \text{if} \quad v_u < dw. \end{cases}$$

In the first case we have

$$S_{u+1} = S_u > \frac{v_0}{(d+1)^u} - 1 > \frac{v_0}{(d+1)^{u+1}} - 1,$$

and in the second case we have

$$S_{u+1} \geq S_u - w + \frac{v_u + w}{d+1} - \frac{d}{d+1}.$$

Since $w \leq S_u$, we have

$$\begin{aligned} S_{u+1} &\geq \frac{S_u}{d+1} + \frac{dw}{d+1} - w + \frac{v_u + w}{d+1} - \frac{d}{d+1} = \frac{S_u}{d+1} + \frac{v_u}{d+1} - \frac{d}{d+1} \\ &\geq \frac{S_u}{d+1} - \frac{d}{d+1} > \left(\frac{v_0}{(d+1)^u} - 1\right)\frac{1}{d+1} - \frac{d}{d+1} = \frac{v_0}{(d+1)^{u+1}} - 1, \end{aligned}$$

and the lemma follows by induction. □

8.5 Proof of Theorem 8.1

Lemma 8.3. *If $s \geq (2k)^{n+1}$, then the congruence*

$$F(\boldsymbol{\lambda}) \equiv 0 \pmod{\wp^{t_0}}$$

has a solution $\boldsymbol{\lambda}$ with a λ_i such that $\wp \not| \alpha_i \lambda_i$.

Proof. Let $k = p^b k_0$, where $p \not| k_0$. Then

$$p^{ef} = N(\wp)^e \leq N(p) = p^n,$$

and so $ef \leq n$ and

$$\begin{aligned} d &= \big(k, N(\wp) - 1\big) = (k, p^f - 1) = (k_0, p^f - 1), \\ d + 1 &\le \min\big(2k_0, p^f\big), \\ (d+1)^{t_0} &\le (d+1)^{\frac{e}{p-1}+be+1} \le p^{\frac{fe}{p-1}+fbe} \cdot (2k_0) \\ &\le 2^{n+1}\big(k_0 p^b\big)^{ef} \le 2^{n+1} k^n. \end{aligned}$$

Therefore, by Lemma 8.2,

$$S_{t_0} > \frac{v_0}{(d+1)^{t_0}} - 1 \ge \frac{s}{k(d+1)^{t_0}} - 1 \ge \frac{s}{(2k)^{n+1)}} - 1 \ge 0.$$

This means that F can be contracted into a form of (8.9) with $u = t_0$, where at least one of the forms $H^{t_0}, H^{t_0+1}, \ldots$ contains a derived variable. We take the derived variable to be 1, and all the other variables to be 0. This gives a solution to $F(\boldsymbol{\lambda}) \equiv 0 \pmod{\wp^{t_0}}$, and on tracing back the derived variable to its ancester in $F^{(0)}$ we see that the solution satisfies the condition stated in the lemma, which is now proved. □

Proof of Theorem 8.1. By Lemmas 6.7 and 8.3, we have, for $\wp \nmid \alpha_i$, $1 \le i \le s$, $s \ge (2k)^{n+1}$ and $t \ge t_0$,

$$M_2(\wp^t) \ge N(\wp)^{(t-t_0)(s-1)}.$$

Set

$$\Lambda = \prod_{i=1}^{s} \alpha_i = \prod_{i=1}^{w} \wp_i^{a_i},$$

where $\wp_i$ are distinct prime ideals. Then, by (8.5), we have, for $s \ge (2k)^{n+1}$, $t \ge t_0 + \big([\frac{a_i}{k}] + 1\big)k$ and $1 \le i \le w$,

$$M(\wp_i^t) \ge M_2\Big(\wp_i^{t-([\frac{a_i}{k}]+1)k}\Big) \ge N(\wp_i)^{\left(t-([\frac{a_i}{k}]+1)k-t_0\right)(s-1)}.$$

By (6.2) we obtain

$$\chi(\wp) = \sum_{i=0}^{\infty} H(\wp^i) \ge \begin{cases} N(\wp)^{-t_0(s-1)}, & \text{if } \wp \nmid \Lambda, \\ N(\wp)^{-\left(([\frac{a}{k}]+1)k+t_0\right)(s-1)}, & \text{if } \wp \| \Lambda. \end{cases}$$

By Theorem 2.1, we have, for $N(\wp) \ge c_5(k, \mathbb{K})$ and $\wp \nmid \Lambda$,

$$\left|\sum_{i=1}^{\infty} H(\wp^i)\right| \ll \sum_{i=1}^{\infty} N(\wp)^{i(1-(2k)^{n+1}/k+\frac{1}{2})} < N(\wp)^{-2},$$

so that

$$\chi(\wp) > 1 - N(\wp)^{-2}.$$

Since

$$t_0 \le n + n\frac{\log k}{\log p} + 1 \le 2n(\log k + 1),$$

we have, for $a \geq 1$,

$$\left(\left[\tfrac{a}{k}\right] + 1\right)k + t_0 \leq a + k + 2n \log k + 2n \leq 2a(n \log k + k + n),$$

and thus, for $\wp^a \| \Lambda$,

$$\chi(\wp) \geq N(\wp^a)^{-2(n \log k + k + n)(s-1)}.$$

Therefore

$$\begin{aligned} \mathfrak{S}(0) = \prod_{\wp} \chi(\wp) &= \prod_{\substack{\wp \nmid \Lambda \\ N(\wp) < c_5}} \chi(\wp) \prod_{\substack{\wp \nmid \Lambda \\ N(\wp) \geq c_5}} \chi(\wp) \prod_{\wp | \Lambda} \chi(\wp) \\ &\geq c_6(k, \mathbb{K}, s) \prod_{N(\wp) \geq c_5} (1 - N(\wp)^{-2}) \prod_{\wp^a \| \Lambda} N(\wp^a)^{-2(n \log k + k + n)(s-1)} \\ &\geq c_7(k, \mathbb{K}, s) N(\Lambda)^{-2(n \log k + k + n)(s-1)}. \end{aligned}$$

The theorem is proved. □

8.6 Proof of Theorem 8.2

In this section we shall assume that k is an odd number exceeding $c_8(\epsilon)$ and we consider the form $F(\boldsymbol{\lambda}) = \sum_{i=1}^{s} \alpha_i \lambda_i^k$ of the type (8.4). In view of the proof of Theorem 8.1, it suffices to prove the following:

Lemma 8.4. *If $s \geq ckn \log k$, then the congruence*

$$F(\boldsymbol{\lambda}) \equiv 0 \pmod{\wp^{t_0}}$$

has a solution with a λ_i such that $\wp \nmid \alpha_i \lambda_i$.

For this lemma we shall require:

Lemma 8.5. *If $\wp \nmid \Lambda$ and $s \geq (\frac{2}{\log 2} + \epsilon) \log k$, then the congruence*

$$F(\boldsymbol{\lambda}) \equiv 0 \pmod{\wp} \tag{8.10}$$

has a nontrivial solution, that is a solution with at least one component not divisible by $\wp$.

Proof. Suppose that $N(\wp) \leq k^{2+\epsilon/2}$. Consider the $2^s - 1$ sums

$$\alpha_i \ (1 \leq i \leq s), \quad \alpha_i + \alpha_j \ (1 \leq i < j \leq s), \quad \ldots, \quad \alpha_i + \cdots + \alpha_s. \tag{8.11}$$

Since

$$2^s - 1 \geq 2^{(\frac{2}{\log 2} + \epsilon) \log k} - 1 = k^{2 + \epsilon \log 2} - 1 > k^{2+\epsilon/2},$$

at least two of the sums are congruent mod $\wp$. Since -1 is a k-th power residue mod $\wp$, the congruence (8.10) has a nontrivial solution.

Suppose now that $N(\wp) > k^{2+\epsilon/2}$. We shall prove that the lemma is still true if $s \geq [\frac{8}{\epsilon}]+4$. Let $M(\wp)$ denote the number of solutions of the congruence

$$\alpha_1\lambda_1^k + \cdots + \alpha_{s-1}\lambda_{s-1}^k \equiv -\alpha_s \pmod{\wp}, \quad \alpha_i \in R(\wp), \quad 1 \leq i \leq s-1.$$

Then, by Lemma 2.5,

$$M(\wp) = N(\wp)^{-1}\sum_{\mu}\sum_{\lambda_1(\wp)}\cdots\sum_{\lambda_{s-1}(\wp)} E\big(\mu(\alpha_1\lambda_1^k + \cdots + \alpha_{s-1}\lambda_{s-1}^k + \alpha_s)\big),$$

where μ runs over a complete residue system of $(\wp\delta)^{-1} \bmod \delta^{-1}$. Therefore

$$M(\wp) = N(\wp)^{s-2} + N(\wp)^{-1}\sum_{\mu}{}^{\star} E(\mu\alpha_s)\prod_{i=1}^{s-1} S(\mu\alpha_i\lambda^k, \wp).$$

By Theorem 2.2,

$$\left|N(\wp)^{-1}\sum_{\mu}{}^{\star} E(\mu\alpha_s)\prod_{i=1}^{s-1} S(\mu\alpha_i\lambda^k, \wp)\right| \leq (k-1)^{s-1}N(\wp)^{\frac{s-1}{2}}.$$

Since

$$\begin{aligned} N(\wp)^{s-2} - (k-1)^{s-1}N(\wp)^{\frac{s-1}{2}} &= N(\wp)^{\frac{s-1}{2}}\left(N(\wp)^{\frac{s-3}{2}} - (k-1)^{s-1}\right) \\ &> N(\wp)^{\frac{s-1}{2}}\left(k^{(2+\epsilon/2)(s-3)/2} - (k-1)^{s-1}\right) \\ &= N(\wp)^{\frac{s-1}{2}}\left(k^{s-3+\frac{\epsilon}{4}(s-3)} - (k-1)^{s-1}\right) \\ &> N(\wp)^{\frac{s-1}{2}}\left(k^{s-1} - (k-1)^{s-1}\right) > 0, \end{aligned}$$

we have $M(\wp) > 0$, that is, (8.10) has a solution $\lambda_1, \ldots, \lambda_{s-1}, 1$, and the lemma is proved. □

Lemma 8.6. *If* $\wp \nmid \Lambda$, $b \geq 1$, $w \geq (b+2)e$ *and* $s \geq (\frac{2}{\log 2}+\epsilon)n\log k$, *then the congruence*

$$F(\boldsymbol{\lambda}) \equiv 0 \pmod{\wp^{w-e}} \tag{8.12}$$

has a nontrivial solution.

Proof. Suppose that $w = (b+2)e$. Since

$$2^s - 1 = k^{2n+n\epsilon\log 2} - 1 > k^{2n} \geq p^{2nb} \geq N(\wp)^{2be} \geq N(\wp)^{(b+1)e},$$

at least two of the sum (8.11) are congruent mod $\wp^{(b+1)e}$; in other words, the congruence (8.12) has a nontrivial solution with $w = (b+2)e$, so that the lemma is established for this case.

Suppose now that $w \geq (b+2)e$ and that (8.12) has a nontrivial solution. We may suppose without loss of generality that $\wp \nmid \lambda_1$. We proceed to show that

$$F(\boldsymbol{\lambda}) \equiv 0 \pmod{\wp^w}$$

also has a nontrivial solution. Set

$$\mu_i = \lambda_i + \nu_i \pi^{w-be-e}, \qquad 1 \leq i \leq s.$$

We begin by showing that

$$\sum_{i=1}^{s} \alpha_i \mu_i^k \equiv \sum_{i=1}^{s} \alpha_i \lambda_i^k + k\pi^{w-be-e} \sum_{i=1}^{s} \alpha_i \nu_i \lambda_i^{k-1} \pmod{\wp^w}. \tag{8.13}$$

In fact, consider the $(h+1)$-th term $\binom{k}{h}\lambda_i^{k-h}(\nu_i\pi^{w-be-e})^h$ in the binomial expansion of

$$\mu_i^k = (\lambda_i + \nu_i \pi^{w-be-e})^k,$$

where $h \geq 2$. Suppose that $\wp^a \| h$. Since $\binom{k}{h} = \frac{k}{h}\binom{k-1}{h-1}$, the $(h+1)$-th term is divisible by $\wp^g$, where

$$g = h(w - be - e) + be - ae.$$

To prove (8.13) it suffices to show that $g \geq w$, or $(h-1)(w-be) \geq he + ae$. Since $w \geq (b+2)e$, it suffices to show that $2(h-1) \geq h + a$, that is

$$h \geq a + 2. \tag{8.14}$$

Since $2 \nmid k$ and $p | k$, we have $3^a \leq p^a \leq h$, and thus (8.14) follows by $3^a = (1+2)^a \geq 1 + 2a \geq a + 2$ for the case $a \geq 1$. If $a = 0$, then (8.14) follows at once from $h \geq 2$.

Write

$$k = \pi^{be}\varphi \qquad \text{and} \qquad \sum_{i=1}^{s} \alpha_i \lambda_i^k = \pi^{w-e}\psi$$

in their $\wp$-adic representations, where $\wp \nmid \varphi$. Then, from (8.13), we have

$$\sum_{i=1}^{s} \alpha_i \mu_i^k \equiv \pi^{w-e}\Big(\psi + \varphi \sum_{i=1}^{s} \alpha_i \nu_i \lambda_i^{k-1}\Big) \pmod{\wp^w}.$$

Since $(\varphi\alpha_1\lambda_1^{k-1}, \wp) = 1$, the congruence

$$\psi + \varphi\alpha_1\lambda_1^{k-1}\nu_1 \equiv 0 \pmod{\wp^e}$$

has a unique solution $\nu_1 \pmod{\wp^e}$. Put $\nu_i = 0, 2 \leq i \leq s$. Then we have a solution $\boldsymbol{\mu}$ of the congruence $F(\boldsymbol{\mu}) \equiv 0 \pmod{\wp^w}$, where $\wp \nmid \mu_1$. The lemma follows by induction. □

Proof of Lemma 8.4. 1) Suppose that $\wp | k$. By Lemma 8.6, it follows that $F^{(0)}(\boldsymbol{\lambda}) \equiv 0 \pmod{\wp^{t_0}}$ has a nontrivial solution if $v_0 \geq cn \log k$, and therefore the lemma is true if $s \geq ckn \log k$.
2) Suppose that $\wp \not| k$. If $v_0 \geq c \log k$ and $w \geq 2$, then, by Lemma 8.5, we may assume that $F^{(0)}(\boldsymbol{\lambda}) \equiv 0 \pmod{\wp^{w-1}}$ has a solution with $\wp \not| \lambda_1$. Now we proceed to show that

$$F^{(0)}(\boldsymbol{\mu}) = \sum_{i=1}^{v_0} \alpha_i \mu_i^k \equiv 0 \pmod{\wp^w} \tag{8.15}$$

also has a nontrivial solution. Set

$$\mu_i = \lambda_i + \nu_i \pi^{w-1}, \qquad 1 \leq i \leq v_0.$$

Then

$$\sum_{i=1}^{v_0} \alpha_i \mu_i^k \equiv \sum_{i=1}^{v_0} \alpha_i \lambda_i^k + k\pi^{w-1} \sum_{i=1}^{s} \alpha_i \nu_i \lambda_i^{k-1} \pmod{\wp^w}.$$

Since the first term on the right hand side of the congruence is divisible by $(k\pi^{w-1}\alpha_1\lambda_1^{k-1}, \wp^w) = \wp^{w-1}$, the congruence

$$\sum_{i=1}^{v_0} \alpha_i \mu_i^k + k\pi^{w-1}\alpha_1\lambda_1^{k-1}\nu_1 \equiv 0 \pmod{\wp^w}$$

has a unique solution $\nu_1 \pmod{\wp}$. Put $\nu_i = 0$, $2 \leq i \leq s$. We obtain a solution $\boldsymbol{\mu}$ of (8.15) with $\wp \not| \mu_1$. It follows by induction that (8.15) has a nontrivial solution for any $w \geq 1$, and in particular for $w = t_0$. Therefore the lemma is true for $s \geq ck \log k$. □

8.7 Bounds for Solutions

Proof of Theorem 8.3. We assume without loss of generality that $\alpha_{s-1}^{(\ell)} < 0$ and $\alpha_s^{(\ell)} > 0$. By Theorem 6.2, the singular series corresponding to equation (8.1) is

$$J(0) = D^{\frac{1}{2}(1-s)} k^{-sn} |N(\alpha_1 \cdots \alpha_s)|^{-\frac{1}{k}} \prod_{\ell} F_\ell \prod_m H_m,$$

where

$$F_\ell = \int_{W_\ell} \prod_{i=1}^{s} w_i^{\frac{1}{k}-1} dw,$$

in which $dw = \prod_{1 \leq i < s} dw_i$, and W_ℓ denotes the domain: $0 \leq w_i \leq |\alpha_i^{(\ell)}|$, $1 \leq i \leq s$ and $\pm w_1 \pm \ldots \pm w_{s-2} - w_{s-1} + w_s = 0$; here the sign before w_i is the sign of $\alpha_i^{(\ell)}$, $1 \leq i \leq s-2$, and where

$$H_m = \int_{V_m} \prod_{i=1}^{s} w_i^{\frac{1}{k}-1} dw\, d\varphi$$

in which $d\varphi = \prod_{1\le i<s} d\varphi_i$ and V_m denotes the domain: $0 \le w_i \le |\alpha_i^{(m)}|^2$, $1 \le i \le s$, $-\pi \le \varphi_j \le \pi$, $1 \le j \le s-1$ and

$$w_s = \left|w_1^{1/2}e^{i\varphi_1} + \cdots + w_{s-1}^{1/2}e^{i\varphi_{s-1}}\right|^2.$$

Let X_ℓ denote the domain:

$$\frac{A^{-n}}{8s} \le w_i \le \frac{A^{-n}}{4s},\ 1 \le i \le s-2, \qquad \frac{A^{-n}}{3} \le w_{s-1} \le \frac{A^{-n}}{2}.$$

Since $|\alpha_j^{(i)}| \ge A^{-n}$, $1 \le i \le n$, $1 \le j \le s$ by (5.9), we have

$$F_\ell \ge c_9(s)A^{(1-\frac{1}{k})n} \int_{X_\ell} \prod_{i=1}^{s-1} w_i^{\frac{1}{k}-1} dw \ge c_{10}(s,n)A^{-\frac{sn}{k}}.$$

Let Y_m be the domain:

$$\frac{A^{-2n}}{64s^2} \le w_i \le \frac{A^{-2n}}{16s^2},\ 1 \le i \le s-2, \qquad \frac{A^{-2n}}{9} \le w_{s-1} \le \frac{A^{-2n}}{4}.$$

Then

$$H_m \ge c_{11}(s)A^{2(1-\frac{1}{k})n} \int_{Y_m} \prod_{i=1}^{s-1} w_i^{\frac{1}{k}-1} dw \ge c_{12}(s,n)A^{-\frac{2sn}{k}}.$$

Thus we have

$$J(0) \ge c_{13}(k,\mathbb{K},s)A^{-(sn+sr_1n+2sr_2n)/k} \le c_{13}(k,\mathbb{K},s)A^{-2sn^2/k}.$$

Let $s_0 = \max\left((2k)^{n+1}, 2^k+1\right)$. If $s \ge s_0$, we may assume that, for each ℓ, the components $\alpha_i^{(\ell)}$ $(1 \le i \le s_0)$ have different signs after a reordering of $\alpha_1, \ldots, \alpha_s$. Since we can put $\lambda_i = 0$, $s_0+1 \le i \le s$, it suffices to prove Theorem 8.3 for the case of $s = s_0$. By Theorem 5.2, the number of solutions of (8.1), with $\lambda_i \in P(T)\,(1 \le i \le s_0)$, is equal to

$$R_{s_0}(0) = \mathfrak{S}(0)J(0)T^{(s_0-k)n} + O\left(\Phi T^{(s_0-k)n}\right),$$

where

$$\Phi = \left(A^{s_0} + A^{2s_0n/k}\right)T^{-a} + A^{2s_0n^2/k}T^{(1-a)n} + T^{-2^{-k}}.$$

Take $c_{14}(k,\mathbb{K})$ sufficiently large, and

$$T = c_{14}A^{2^{k+1}s_0^2(n\log k+k+n)n}.$$

Since

$$\mathfrak{S}(0)J(0) > c_{15}(k,\mathbb{K})A^{-2s_0(s_0-1)(n\log k+k+n)n-2s_0n^2/k},$$

we have

$$R_{s_0}(0) > 1,$$

and Theorem 8.3 is proved with

$$c_3 = \max\left((2k)^{n+1}, 2^k + 1\right) \qquad \text{and} \qquad c_4 = c_3^2 2^{k+1}(n \log k + k + n)n.$$

If $r_1 = 0$, we have $\prod_\ell F_\ell = 1$. Therefore the above argument yields Theorem 8.4. Finally suppose that $2 \nmid k$. Consider the equation

$$B(\boldsymbol{\mu}) = \sum_{i=1}^{s} \beta_i \mu_i^k = 0 \tag{8.16}$$

where

$$\begin{aligned} \beta_{2\ell-1} &= |\alpha_{2\ell-1}^{(\ell)}| \, \alpha_{2\ell-1}^{(\ell)-1} \alpha_{2\ell-1}, && 1 \le \ell \le r_1, \\ \beta_{2\ell} &= -|\alpha_{2\ell}^{(\ell)}| \, \alpha_{2\ell}^{(\ell)-1} \alpha_{2\ell}, && 1 \le \ell \le r_1, \\ \beta_i &= \alpha_i, && 2r_1 \le i \le s. \end{aligned}$$

Then $\beta_{2\ell-1}^{(\ell)}$ and $\beta_{2\ell}^{(\ell)}$ have different signs, and so, by Theorem 8.3, the equation (8.16) has a nontrivial solution in P^s having

$$\|\boldsymbol{\mu}\| \ll \max_i \|\beta_i\|^{c_4(k,n)} = A^{c_4(k,n)}.$$

Consequently equation (8.1) has a solution $\boldsymbol{\lambda} \neq \mathbf{0}$ in J^s satisfying

$$\|\boldsymbol{\lambda}\| \ll A^{c_4(k,n)},$$

where

$$\begin{aligned} \lambda_{2\ell-1} &= |\alpha_{2\ell-1}^{(\ell)}| \, \alpha_{2\ell-1}^{(\ell)-1} \mu_{2\ell-1}, && 1 \le \ell \le r_1, \\ \lambda_{2\ell} &= -|\alpha_{2\ell}^{(\ell)}| \, \alpha_{2\ell}^{(\ell)-1} \mu_{2\ell}, && 1 \le \ell \le r_1, \\ \lambda_i &= \mu_i, && 2r_1 + 1 \le i \le s. \end{aligned}$$

Theorem 8.5 is proved. □

Notes

Peck [1] first proved that $\mathfrak{S}(0) > 0$ if $s \ge 4k^{2n+3} + 1$. However, for the case $\mathbb{K} = \mathbb{Q}$, Davenport and Lewis [1] proved that $\mathfrak{S}(0) > 0$, if $s \ge k^2 + 1$ with equality whenever $k + 1$ is a prime, and Chowla and Shimula [1] show that $\mathfrak{S}(0) > 0$ if k is odd and $s \ge ck \log k$, which is best possible apart from improvements on the constant c. Their methods can be generalised to treat the singular series for additive equations in $\mathbb{K}$, and Theorems 8.1 and 8.2 were obtained by Wang [4].

Theorem 8.3 is a generalisation of a theorem due to Pitman [1]. She first established the conclusion of Theorem 8.3 for the case $\mathbb{K} = \mathbb{Q}$ with $c_3 = c_4 = ck^2 \log k$.

Lemma 8.1: See Chevalley [1].

Chapter 9

Small Nonnegative Solutions of Additive Equations

9.1 Introduction

Let $\alpha_1, \dots, \alpha_{2s}$ be $2s$ nonzero elements of P. Consider the additive equation of the type

$$\alpha_1 \lambda_1^k + \cdots + \alpha_s \lambda_s^k - \alpha_{s+1} \lambda_{s+1}^k - \cdots - \alpha_{2s} \lambda_{2s}^k = 0. \tag{9.1}$$

A set of numbers $\lambda_1, \dots, \lambda_{2s}$ satisfying (9.1) is called a nontrivial solution of (9.1) if $\lambda_i \in P\,(1 \le i \le 2s)$, not all zero. Set

$$M = \max_i N(\alpha_i).$$

Theorem 9.1. *Suppose that $s \ge c_1(k, n, \epsilon)$. Then equation (9.1) has a nontrivial solution such that*

$$\max_i N(\lambda_i) \ll M^{\frac{1}{k}+\epsilon}. \tag{9.2}$$

A key point in the proof of this theorem is the reduction of (9.1) to the equation

$$\alpha_1 \lambda_1^k + \cdots + \alpha_s \lambda_s^k - \alpha_{s+1} \lambda_{s+1}^k - \cdots - \alpha_{2s} \lambda_{2s}^k = d\chi$$

of $\lambda_1, \dots, \lambda_{2s}, \chi$, where $d = \pm 1$. This latter equation, which contains a linear term $d\chi$, is much easier to handle than (9.1).

9.2 Hurwitz's Lemma

Lemma 9.1 (Hurwitz). *There exists a rational integer $c_2 = c_2(\mathbb{K})$ with the following property. Corresponding to any α, $\beta\,(\neq 0)$ of J, there is a positive integer $q \le c_2$, and an integer w of J satisfying $|N(q\alpha - w\beta)| < |N(\beta)|$.*

Proof. Let $\gamma = \alpha/\beta$. It suffices to show that there is a c_2 with the property that, for any $\gamma \in \mathbb{K}$, we have $|N(q\gamma - w)| < 1$ for some $q \le c_2$ and some $w \in J$. Write $\gamma = \sum_{1 \le i \le n} r_i w_i$ with $r_i \in \mathbb{Q}$, $1 \le i \le n$. Then

$$|N(\gamma)| = \Big|\prod_i \Big(\sum_j r_j w_j^{(i)}\Big)\Big| \le c_3 \big(\max_i |r_i|\big)^n,$$

where $c_3 = \prod_i (\sum_j |w_j^{(i)}|)$. Choose an integer $t > \sqrt[n]{c_3}$ and set $c_2 = t^n$. Write $r_i = a_i + b_i$, where $a_i \in \mathbb{Z}$ and $0 \le b_i < 1$, $1 \le i \le n$. Let

$$[\gamma] = \sum_{i=1}^n a_i w_i \qquad \text{and} \qquad \{\gamma\} = \sum_{i=1}^n b_i w_i.$$

Then $\gamma = [\gamma] + \{\gamma\}$, where $[\gamma] \in J$. Map $\mathbb{K}$ to E_n by

$$\phi\Big(\sum_{i=1}^n r_i w_i\Big) = (r_1, \ldots, r_n).$$

For any $\gamma \in \mathbb{K}$, the point $\phi(\{\gamma\})$ lies in the unit cube. Partition the unit cube into t^n subcubes with side $1/t$. Consider the points $\phi(\{q\gamma\})$ for $1 \le q \le t^n + 1$. By the box principle of Dirichlet, at least two of them must lie in the same subcube, say those corresponding to $q_1\gamma$ and $q_2\gamma$. If we write $q_1\gamma = [q_1\gamma] + \{q_1\gamma\}$ and $q_2\gamma = [q_2\gamma] + \{q_2\gamma\}$, and subtract, then we find that $q\gamma = w + \tau$, where (assuming that $q_1 > q_2$) $q = q_1 - q_2 \le t^n$, $w \in J$ and the coordinates of τ have absolute values at most $1/t$. Thus $N(\tau) \le c_3(\frac{1}{t})^n < 1$. The lemma is proved. □

The above lemma is a weak generalisation of the Euclidean algorithm in an algebraic number field.

Lemma 9.2. *For any t integers $\gamma_1, \ldots, \gamma_t$, not all zero, let γ be a nonzero element of the integral ideal $\mathfrak{a} = (\gamma_1, \ldots, \gamma_t)$ with the least norm in absolute value. Then there exists a rational integer $c_4(\mathbb{K})$ such that*

$$c_4\gamma_i \big/ \gamma, \qquad 1 \le i \le t$$

are integers.

Proof. Set $\alpha = \gamma_i$ and $\beta = \gamma$ in Lemma 9.1. Then there exist a rational integer q_i and an integer σ_i such that

$$|N(q_i\gamma_i - \sigma_i\gamma)| < |N(\gamma)|, \qquad 1 \le q \le c_2.$$

Since $q_i\gamma_i - \sigma_i\gamma \in \mathfrak{a}$, it follows that $q_i\gamma_i - \sigma_i\gamma = 0$; that is $q_i\gamma_i/\gamma$ is an integer. Set $c_4 = c_2!$. Then $q_i | c_4$ and

$$c_4\gamma_i \big/ \gamma = \frac{c_4}{q_i}\Big(\frac{q_i\gamma_i}{\gamma}\Big), \qquad 1 \le i \le t$$

are integers. The lemma is proved. □

Lemma 9.3. *For any t vectors (γ_i, δ_i), $(1 \le i \le t)$ of J^2, where $\gamma_i \ne 0$ for $1 \le i \le t$, if $\delta_1/\gamma_1 = \cdots = \delta_t/\gamma_t$, then*

$$(\gamma_i, \delta_i) = c_4^{-1}\chi_i(\gamma, \delta), \qquad 1 \le i \le t,$$

where γ is defined in Lemma 9.2, *and where δ and $\chi_i\,(1 \le i \le t)$ are integers.*

Proof. By Lemma 9.2, $\chi_i = c_4\gamma_i/\gamma$ $(1 \le i \le t)$ are integers. Let $\delta_1/\gamma_1 = \cdots = \delta_t/\gamma_t = \sigma$. Then $\delta_i = \sigma\gamma_i$, $1 \le i \le t$. Since $(\delta_1, \ldots, \delta_t) = \sigma(\gamma_1, \ldots, \gamma_t) = \sigma\mathfrak{a}$ is an integral ideal, and $\gamma \in \mathfrak{a}$, $\delta = \sigma\gamma$ is an integer, it follows that

$$c_4^{-1}\chi_i\delta = c_4^{-1}\chi_i\sigma\gamma = c_4^{-1}\chi_i\delta_i\gamma_i^{-1}\gamma = c_4^{-1}\chi_i\delta_i\gamma(c_4^{-1}\chi_i\gamma)^{-1} = \delta_i, \quad 1 \le i \le t.$$

The lemma is proved. □

9.3 Reductions

Proposition 9.1. *Suppose that $x \ge 1/k$ and $s \ge c_5(k, n, x, \epsilon)$. Then* (9.1) *has a nontrivial solution with*

$$\max_i N(\lambda_i) \ll M^{x+\epsilon}.$$

The case $x = 1/k$ is Theorem 9.1. From Theorem 8.3 we find that if $s \ge c_6(k, n)$, then the equation of the type

$$\alpha_1\lambda_1^k + \cdots + \alpha_t\lambda_t^k - \alpha_{t+1}\lambda_{t+1}^k - \cdots - \alpha_s\lambda_s^k = 0$$

has a solution in integers $\lambda_1, \ldots, \lambda_s$ of P, not all zero and such that

$$\max_i N(\lambda_i) \ll \max_i N(\alpha_i)^{c_7(k,n)}, \tag{9.3}$$

where α_i are given nonzero integers of P and $1 \le t \le s-1$.

It suffices to prove the proposition when M is large, say $M \ge c_8(k, \mathbb{K}, x, \epsilon)$. In fact, if $M < c_8$ and $s \ge c_6$, then it follows from (9.3) that (9.1) has a nontrivial solution such that

$$\max_i N(\lambda_i) \ll M^{c_7} \ll c_8^{c_7} \ll M^{x+\epsilon}.$$

Let X be the set of x such that Proposition 9.1 holds. Then (9.3) shows that X is not empty, and it is clear that X is a closed set. Hence the proof of Proposition 9.1 is reduced to proving that if $x > 1/k$ and $x \in X$, then there exists an $x' \in X$ with $x' < x$.

By Lemma 3.1 there exists a set of totally nonnegative units σ_i, $1 \le i \le 2s$ such that

$$c_9^{-1}N(\alpha_i)^{\frac{1}{n}} < \alpha_i^{(\ell)}\sigma_i^{(\ell)k} < c_9 N(\alpha_i)^{\frac{1}{n}}, \quad c_9^{-1}N(\alpha_i)^{\frac{1}{n}} < |\alpha_i^{(m)}\sigma_i^{(m)k}| < c_9 N(\alpha_i)^{\frac{1}{n}},$$

for $1 \leq i \leq 2s$, with $c_9 = c_9(k, \mathbb{K})$. Let $\alpha_i' = \alpha_i \sigma_i^k$, $\lambda_i = \sigma_i \lambda_i'$, $1 \leq i \leq 2s$. Then (9.1) becomes

$$\alpha_1' {\lambda_1'}^k + \cdots + \alpha_s' {\lambda_s'}^k - \alpha_{s+1}' {\lambda_{s+1}'}^k - \cdots - \alpha_{2s}' {\lambda_{2s}'}^k = 0. \tag{9.4}$$

If the proposition holds for x' and for the particular equation (9.4), then it has a nontrivial solution of (9.4) such that

$$\max_i N(\lambda_i') \ll \max_i N(\alpha_i')^{x'+\epsilon}.$$

Since $N(\lambda_i') = N(\lambda_i)$, $N(\alpha_i') = N(\alpha_i)$, $1 \leq i \leq 2s$, we have a nontrivial solution of (9.1) with $\max_i N(\lambda_i) \ll M^{x'+\epsilon}$; thus Propsoition 9.1 holds for x', and for (9.1). Hence we may suppose without loss of generality that

$$c_9^{-1}N(\alpha_i)^{\frac{1}{n}} < \alpha_i^{(\ell)} < c_9 N(\alpha_i)^{\frac{1}{n}}, \quad c_9^{-1}N(\alpha_i)^{\frac{1}{n}} < |\alpha_i^{(m)}| < c_9 N(\alpha_i)^{\frac{1}{n}} \tag{9.5}$$

hold for $1 \leq i \leq 2s$.

In what follows x will be a fixed number exceeding $1/k$ for which Proposition 9.1 holds. Take a sufficiently small y such that

$$\frac{1}{k} + 6c_7 ny + 20ny < x \qquad \text{and} \qquad 22kny < 1, \tag{9.6}$$

and put

$$x' = \max\left(x(1 - \tfrac{1}{2}y) + \frac{y}{2kn}, \frac{1}{k} + 6c_7 ny + 20ny\right), \tag{9.7}$$

so that $x' < x$. We proceed to prove that Proposition 9.1 holds for x'.

Let $\epsilon_1 = \min(\frac{\epsilon}{8x'}, \frac{\epsilon}{4})$, and divide the interval $[0,1]$ into a finite number of intervals I of length $\leq \epsilon_1$. If s is large, one of these intervals I will have the property that there are more than $[s/\epsilon_1]$ numbers among $\alpha_1, \ldots, \alpha_s$ that are of the type

$$N(\alpha_i) = M^{a_i}, \qquad a_i \in I.$$

We may also suppose that

$$\frac{N(\alpha_i)}{N(\alpha_j)} \leq M^{\epsilon_1}, \qquad 1 \leq i, j \leq s,$$

and that the above relation holds for $s+1 \leq i, j \leq 2s$. Let $a^n = M^{\epsilon_1} \max_{1 \leq i \leq s} N(\alpha_i)$ and $b^n = M^{\epsilon_1} \max_{1 \leq i \leq s} N(\alpha_{s+i})$. Let p_i and q_i be the largest rational integers such that

$$N(\alpha_i) p_i^{kn} \leq a^n \quad \text{and} \quad N(\alpha_{s+i}) q_i^{kn} \leq b^n, \qquad 1 \leq i \leq s.$$

Since $M \geq c_8$ and

$$\frac{a^n}{N(\alpha_i)} \geq M^{\epsilon_1}, \qquad \frac{b^n}{N(\alpha_{s+i})} \geq M^{\epsilon_1}$$

we may suppose that

$$p_i \geq 2^{-\frac{1}{kn}} \left(\frac{a^n}{N(\alpha_i)} \right)^{\frac{1}{kn}} \quad \text{and} \quad q_i \geq 2^{-\frac{1}{kn}} \left(\frac{b^n}{N(\alpha_{s+i})} \right)^{\frac{1}{kn}} \qquad 1 \leq i \leq s.$$

Hence

$$N(\alpha_i) p_i^{kn} \geq \frac{a^n}{2} \quad \text{and} \quad N(\alpha_{s+i}) q_i^{kn} \geq \frac{b^n}{2} \qquad 1 \leq i \leq s.$$

Set $\alpha'_i = \alpha_i p_i^k$, $\alpha'_{s+i} = \alpha_{s+i} q_i^k$, $\lambda = p_i \lambda'_i$, $\lambda_{s+1} = q_i \lambda'_{s+i}$, $1 \leq i \leq s$. Then (9.1) becomes (9.4), and by (9.5), $\alpha'_i \, (1 \leq i \leq 2s)$ satisfy

$$\begin{aligned} &(2c_9)^{-1} a < {\alpha'_i}^{(\ell)} < c_9 a, \qquad &(2c_9)^{-1} a < |{\alpha'_i}^{(m)}| < c_9 a, \qquad &1 \leq i \leq s, \\ &(2c_9)^{-1} b < {\alpha'_{s+i}}^{(\ell)} < c_9 b, \qquad &(2c_9)^{-1} b < |{\alpha'_{s+i}}^{(m)}| < c_9 b, \qquad &1 \leq i \leq s. \end{aligned}$$

Suppose that Proposition 9.1 holds for x' and for the particular equation (9.4). Then there exists a nontrivial solution of (9.4) with

$$\max_i N(\lambda'_i) \ll \max(a^n, b^n)^{x' + \epsilon/4} \ll M^{(1+\epsilon_1)(x' + \epsilon/4)} \ll M^{x' + \epsilon/2}.$$

Since

$$\begin{aligned} N(\alpha_i) &= M^{\epsilon_1} N(\alpha_i) \max_{1 \leq j \leq s} N(\alpha_j) \Big/ M^{\epsilon_1} \max_{1 \leq q \leq s} N(\alpha_q) \\ &= a^n M^{-\epsilon_1} N(\alpha_i) \Big/ \max_{1 \leq q \leq s} N(\alpha_q) \geq a^n M^{-2\epsilon_1}, \quad 1 \leq i \leq s, \end{aligned}$$

we have

$$p_i^n \leq p_i^{kn} \leq \frac{a^n}{N(\alpha_i)} \leq M^{2\epsilon_1} \leq M^{\epsilon/2}, \qquad 1 \leq i \leq s,$$

and therefore

$$N(\lambda_i) \ll p_i^n N(\lambda'_i) \ll M^{x' + \epsilon}, \qquad 1 \leq i \leq s.$$

Similarly

$$N(\lambda_{s+i}) \ll M^{x' + \epsilon}, \qquad 1 \leq i \leq s.$$

Thus Proposition 9.1 holds for x' and for (9.1), so that, in proving that proposition for x', we may suppose that

$$\begin{aligned} &c_{10} a < \alpha_i^{(\ell)} < c_{11} a, \qquad &c_{10} a < |\alpha_i^{(m)}| < c_{11} a, \qquad &1 \leq i \leq s, \\ &c_{10} b < \alpha_{s+i}^{(\ell)} < c_{11} b, \qquad &c_{10} b < |\alpha_{s+i}^{(m)}| < c_{11} b, \qquad &1 \leq i \leq s, \end{aligned} \tag{9.8}$$

for certain numbers a, b and where $c_{10} = c_{10}(k, \mathbb{K})$ and $c_{11} = c_{11}(k, \mathbb{K})$.

9.4 Continuation

In what follows, q will be the integer $c_5(k,n,x,\epsilon)$ occuring in Propsoition 9.1, and $s > q$. Set

$$z = \frac{y}{2kn^2}. \tag{9.9}$$

We distinguish two cases.

A. There is a subset of q elements among $\alpha_1, \ldots, \alpha_s$, say $\alpha_1, \ldots, \alpha_q$, and there is a subset of q elements among $\alpha_{s+1}, \ldots, \alpha_{2s}$, say $\alpha_{s+1}, \ldots, \alpha_{s+q}$, and there are integers $\sigma_1, \ldots, \sigma_q, \sigma_{s+1}, \ldots, \sigma_{s+q}$ of P such that

$$0 < \|\sigma_i\| \le M^z, \qquad 1 \le i \le q \quad \text{or} \quad s+1 \le i \le s+q \tag{9.10}$$

and

$$|N(\sigma)| \ge M^y,$$

where σ is a nonzero element in the integral ideal $(\alpha_1\sigma_1, \ldots, \alpha_q\sigma_q, \alpha_{s+1}\sigma_{s+1}, \ldots, \alpha_{s+q}\sigma_{s+q})$ with the least norm in absolute value.

By Lemma 3.10 we may choose a nonzero integer γ such that $\|\gamma\| \le c_{12}(\mathbb{K})$ and $\gamma\sigma \in P$. By Lemma 9.2,

$$\alpha_i' = \frac{c_4\alpha_i\sigma_i^k\gamma^2}{\gamma\sigma}, \qquad 1 \le i \le q \quad \text{or} \quad s+1 \le i \le s+q,$$

are all totally nonnegative integers. Therefore it follows from the case x of Proposition 9.1 that the equation

$$\alpha_1'{\lambda_1'}^k + \cdots + \alpha_q'{\lambda_q'}^k - \alpha_{s+1}'{\lambda_{s+1}'}^k + \cdots - \alpha_{s+q}'{\lambda_{s+q}'}^k = 0$$

has a nontrivial solution satisfying

$$\max_{\substack{1 \le i \le q \\ s+1 \le i \le s+q}} N(\lambda_i') \ll \max_{\substack{1 \le i \le q \\ s+1 \le i \le s+q}} N(\alpha_i')^{x+\epsilon} \ll M^{(1+knz-y)(x+\epsilon)}.$$

Set $\lambda_i = \sigma_i\lambda_i'$, $\lambda_{s+i} = \sigma_{s+i}\lambda_{s+i}'$, $1 \le i \le q$ and $\lambda_j = 0$, $q < j \le s$ or $s+q < j \le 2s$. Then, by (9.7) and (9.9), equation (9.1) has a nontrivial solution satisfying

$$\max N(\lambda_i) \ll M^{(1+knz-y)(x+\epsilon)+nz} \ll M^{(1-y/2)(x+\epsilon)+nz} \ll M^{x'+\epsilon}.$$

We are thus reduced to case:

B. For any q elements, say $\alpha_1, \ldots, \alpha_q$ among $\alpha_1, \ldots, \alpha_s$, and for any q elements, say $\alpha_{s+1}, \ldots, \alpha_{s+q}$ among $\alpha_{s+1}, \ldots, \alpha_{2s}$, and given any $2q$ integers $\sigma_1, \ldots, \sigma_q, \sigma_{s+1}, \ldots, \sigma_{s+q}$ of P satisfying (9.10), the integer σ defined in case A satisfies $|N(\sigma)| < M^y$.

Condition B depends on k, n, q, M, y and will be denoted by $B(k,n,q,M,y)$.

Proposition 9.2. *Let* $d = 1$ *or* -1,

$$M = \max\left(a^n, b^n\right), \tag{9.11}$$

and let $\alpha_1, \ldots, \alpha_{2s}$ *be* $2s$ *nonzero integers of* P *satisfying* (9.8) *and* $B(k,n,q,M,y)$. *Then, for* $s \geq c_{13}(k,n,q,y)$, *the equation*

$$\alpha_1\lambda_1^k + \cdots + \alpha_s\lambda_s^k - \alpha_{s+1}\lambda_{s+1}^k + \cdots - \alpha_{2s}\lambda_{2s}^k = d\chi$$

has a solution in integers $\lambda_1, \ldots, \lambda_{2s}, \chi$ *of* P, *not all zero and satisfying*

$$\max_i N(\lambda_i) \ll M^{\frac{1}{k}+20ny}, \qquad \|\chi\| \leq M^{6y}.$$

Now we proceed to show that Proposition 9.2 implies that Proposition 9.1 is true for x'. Let x, x', y, z, q be as above. Suppose that c_6 and c_{13} are integers. Let $s = uv$ where $u = c_{13}$ and $v = c_6$. Replace the indices $1 \leq i \leq 2s$ by double indices $1 \leq i \leq v$, $1 \leq j \leq 2u$. Then equation (9.1) can be written as

$$\sum_{i=1}^{v}\left(\alpha_{i1}\lambda_{i1}^k + \cdots + \alpha_{iu}\lambda_{iu}^k - \alpha_{i,u+1}\lambda_{i,u+1}^k - \cdots - \alpha_{i,2u}\lambda_{i,2u}^k\right) = 0. \tag{9.12}$$

For each i, $1 \leq i \leq v$, the coefficients $\alpha_{i1}, \ldots, \alpha_{iu}, \alpha_{i,u+1}, \ldots, \alpha_{i,2u}$ satisfy the conditions in Proposition 9.2. Hence there are integers $\lambda'_{i1}, \ldots, \lambda'_{iu}, \lambda'_{i,u+1}, \ldots, \lambda'_{i,2u}, \chi$ of P, not all zero, such that

$$\alpha_{i1}\lambda_{i1}'^k + \cdots + \alpha_{iu}\lambda_{iu}'^k - \alpha_{i,u+1}\lambda_{i,u+1}'^k + \cdots - \alpha_{i,2u}\lambda_{i,2u}'^k = d_i\chi_i$$

with

$$\max_{i,j} N(\lambda'_{ij}) \ll M^{\frac{1}{k}+20ny}, \qquad \|\chi_i\| \leq M^{6y}.$$

We may suppose that $\chi_i \neq 0$, $1 \leq i \leq v$, since otherwise we get small solutions straightaway. Take $d_1 = \cdots = d_{v-1} = 1$ and $d_v = -1$. Then, by (9.3), the equation

$$\chi_1\gamma_1^k + \cdots + \chi_{v-1}\gamma_{v-1}^k - \chi_v\gamma_v^k = 0$$

has a nontrivial solution satisfying

$$\max_i N(\gamma_i) \ll M^{6c_7ny}.$$

Let $\lambda_{ij} = \gamma_i\lambda'_{ij}$, $1 \leq i \leq v$, $1 \leq j \leq 2u$. Then we have a nontrivial solution of (9.12), that is (9.1), having

$$\max_{i,j} N(\lambda_{ij}) \ll M^{\frac{1}{k}+20ny+6c_7ny} \ll M^{x'}$$

by (9.7). Thus Proposition 9.1 holds for x'.

9.5 Farey Division

Let

$$h = abM^{20ky-3y/2n} \quad \text{and} \quad t = M^{y/n} \tag{9.13}$$

We take the Farey division of U_n with respect to (h,t) and define $\Gamma(t), B_\gamma, B$ and S as in §5.1. Let B'_γ denote the set

$$\{\mathbf{x} : \mathbf{x} \in U_n,\ h\|\xi - \gamma_0\| \le t^{-1} \le N(\mathfrak{a})^{-1/n} \text{ for some } \gamma_0 \equiv \gamma \pmod{\delta^{-1}}\}.$$

Then $B'_\gamma \subset B_\gamma (\gamma \in \Gamma(t))$ and it follows from Lemma 5.1 that if γ_1 and γ_2 belong to $\Gamma(t)$ and $\gamma_1 \ne \gamma_2$, then $B'_{\gamma_1} \cap B'_{\gamma_2} = \phi$. Set

$$B' = \bigcup_{\gamma \in \Gamma(t)} B'_\gamma$$

and define the supplementary domain S' of B' with respect to U_n by

$$S' = U_n - B'.$$

Then $S' \supset S$.

We use the following notations:

$$T_1 = b^{1/k} M^{20y}, \quad T_1 = a^{1/k} M^{20y}, \quad H = M^{6y},$$

$$\begin{cases} S_i(\xi, T_1) = \sum_{\lambda \in P(T_1)} E(\alpha_i \lambda^k \xi), & 1 \le i \le s, \\ S_{s+i}(\xi, T_2) = \sum_{\lambda \in P(T_2)} E(-\alpha_{s+i} \lambda^k \xi), & 1 \le i \le s, \\ S(\xi, T) = \prod_{i=1}^{s} S_i(\xi, T_1) S_{s+i}(\xi, T_2), \end{cases} \tag{9.14}$$

and

$$F(\xi) = \sum_{\chi \in P(H)} S(\xi, T) E(-d\chi\xi),$$

where $d = 1$ or -1. Let R be the number of solutions of the equation

$$\alpha_1 \lambda_1^k + \cdots + \alpha_s \lambda_s^k - \alpha_{s+1} \lambda_{s+1}^k - \cdots - \alpha_{2s} \lambda_{2s}^k = d\chi$$

in integers $\lambda_1, \ldots, \lambda_{2s}, \chi$ satisfying

$$\lambda_i \in P(T_1), \quad \lambda_{s+i} \in P(T_2), \quad 1 \le i \le s, \qquad \chi \in P(H).$$

Then

$$R = \int_{B'} F(\xi)\, dx + \int_{S'} F(\xi)\, dx. \tag{9.15}$$

We shall show that, under the assumption made in Proposition 9.2, this number R is greater than 1.

9.6 The Supplementary Domain

Let

$$\epsilon_2 < \frac{1}{2G}, \qquad \epsilon_2(1+20y) < \tfrac{1}{2}z, \tag{9.16}$$

and

$$s > \frac{10G}{z} + q. \tag{9.17}$$

Lemma 9.4. *Suppose that* $\mathbf{x} \in U_n$ *and that*

$$|F(\xi)| \gg H^n (T_1 T_2)^{sn} M^{-4}. \tag{9.18}$$

Then $\mathbf{x} \in B'$.

Proof. We may suppose that

$$|S_1(\xi, T_1)| \geq \cdots \geq |S_s(\xi, T_1)|.$$

Then

$$F(\xi) \ll H^n T_1^{(q-1)n} T_2^{sn} |S_q(\xi, T_1)|^{s-q+1},$$

and thus, by (9.18) and $M \geq c_8$, we have

$$|S_i(\xi, T_1)| \geq |S_q(\xi, T_1)| \geq T_1^n M^{\frac{-5}{s-q+1}} = C,$$

say, for $1 \leq i \leq q$. By (9.14), (9.16) and (9.17), we have

$$M^{\frac{5}{s-q+1}} \leq T_1^{\frac{1}{4y(s-q+1)}} \leq T_1^{\frac{1}{2G}} < T_1^{\frac{1}{G}-\epsilon_2},$$

and therefore

$$C \geq T_1^{n-\frac{1}{G}+\epsilon_2}.$$

It follows from Theorem 3.3 that there are integers $\sigma_i \, (1 \leq i \leq q)$ of P and $\varphi_i \, (1 \leq i \leq q)$ of J such that

$$0 < \|\sigma_i\| \ll M^{\frac{5G}{s-q+1}} T_1^{\epsilon_2} < M^{\frac{z}{2}+\frac{z}{2}} = M^z$$

and

$$\|\xi \alpha_i \sigma_i - \varphi_i\| \ll M^{\frac{5G}{s-q+1}} T_1^{\epsilon_2 - k} < M^z T_1^{-k}, \quad 1 \leq i \leq q,$$

since $M \geq c_8$. After a reordering of $\alpha_{s+1}, \ldots, \alpha_{2s}$, we may also suppose that

$$|S_{s+1}(\xi, T_2)| \geq \cdots \geq |S_{2s}(\xi, T_2)|.$$

Similarly there are integers $\sigma_{s+i}\,(1 \leq i \leq q)$ of P and integers $\varphi_{s+i}\,(1 \leq i \leq q)$ of J having

$$0 < \|\sigma_{s+i}\| < M^z \quad \text{and} \quad \|\xi\alpha_{s+i}\sigma_{s+i} - \varphi_{s+i}\| < M^z T_2^{-k}, \quad 1 \leq i \leq q.$$

Hence, by (9.8), (9.9), (9.14) and $M \geq c_8$, we have

$$\begin{aligned}
&\|\varphi_i\alpha_{s+j}\sigma_{s+j} - \varphi_{s+j}\alpha_i\sigma_i\| \\
&\qquad = \|\varphi_i\alpha_{s+j}\sigma_{s+j} - \xi\alpha_i\sigma_i\alpha_{s+j}\sigma_{s+j} + \xi\alpha_i\sigma_i\alpha_{s+j}\sigma_{s+j} - \varphi_{s+j}\alpha_i\sigma_i\| \\
&\qquad \leq \|\alpha_{s+j}\sigma_{s+j}\|\,\|\xi\alpha_i\sigma_i - \varphi_i\| + \|\alpha_i\sigma_i\|\,\|\xi\alpha_{s+j}\sigma_{s+j} - \varphi_{s+j}\| \\
&\qquad \ll bM^{2z}T_1^{-k} + aM^{2z}T_2^{-k} \ll M^{2z-20ky} < 1, \qquad 1 \leq i,j \leq q,
\end{aligned}$$

and thus

$$N(\varphi_i\alpha_{s+j}\sigma_{s+j} - \varphi_{s+j}\alpha_i\sigma_i) = 0.$$

Since the argument $\varphi_i\alpha_{s+j}\sigma_{s+j} - \varphi_{s+j}\alpha_i\sigma_i$ is an integer, it is zero. Thus, by Lemma 9.3, the $2q$ vectors $c_4(\alpha_i\sigma_i, \varphi_i)$, $1 \leq i \leq q$ and $s+1 \leq i \leq s+q$ are all integral multiples of an integral vector (σ, τ), where σ is a nonzero element of the integral ideal $(\alpha_1\sigma_1, \ldots, \alpha_q\sigma_q, \alpha_{s+1}\sigma_{s+1}, \ldots, \alpha_{s+q}\sigma_{s+q})$ with the least norm in absolute value. Therefore the condition in case B yields that

$$0 < |N(\sigma)| < M^y.$$

Let

$$\sigma^{-1}\tau\delta = \frac{\mathfrak{b}}{\mathfrak{a}}, \qquad (\mathfrak{a}, \mathfrak{b}) = 1.$$

Then $\mathfrak{a}|\sigma$, and thus

$$N(\mathfrak{a}) \leq |N(\sigma)| < M^y = t^n.$$

Since $\|\sigma_1\| < M^z$, we have, by (5.9),

$$|\sigma_1^{(i)}|^{-1} \leq M^{(n-1)z}, \qquad 1 \leq i \leq n,$$

and, by (9.8), (9.9), (9.13) and (9.14),

$$\begin{aligned}
|\xi^{(i)} - \sigma^{(i)-1}\tau^{(i)}| &= |\alpha_1^{(i)}\sigma_1^{(i)}|^{-1}|\xi^{(i)}\alpha_1^{(i)}\sigma_1^{(i)} - \varphi_1^{(i)}| \\
&\ll a^{-1}M^{nz}T_1^{-k} = a^{-1}b^{-1}M^{-20ky+nz} \\
&< (ht)^{-1}, \qquad 1 \leq i \leq n.
\end{aligned}$$

Therefore $\xi \in B'_\gamma$ where $\gamma \equiv \sigma^{-1}\tau \pmod{\delta^{-1}}$. The lemma is proved. □

9.7 Basic Domains

We use the following notations:

$$\xi - \gamma = \zeta,$$
$$G_i(\gamma) = N(\mathfrak{a}_i)^{-1} \sum_{\lambda(\mathfrak{a}_i)} E(\alpha_i \lambda^k \gamma),$$
$$G_{s+i}(\gamma) = N(\mathfrak{a}_{s+i})^{-1} \sum_{\lambda(\mathfrak{a}_{s+i})} E(-\alpha_{s+i} \lambda^k \gamma),$$
$$I_i(\zeta, T) = \int_0^T E(\alpha_i \eta^k \zeta)\, dy,$$
$$I_{s+i}(\zeta, T) = \int_0^T E(-\alpha_{s+i} \eta^k \zeta)\, dy,$$
$$G(\gamma) = \prod_{i=1}^{2s} G_i(\gamma),$$
$$I(\zeta, T) = \prod_{i=1}^{s} I_i(\zeta, T_1) I_{s+i}(\zeta, T_2),$$
$$I_i(\zeta) = \int_0^1 E(\gamma_i \eta^k \zeta)\, dy,$$
$$I_{s+i}(\zeta) = \int_0^1 E(-\gamma_{s+i} \eta^k \zeta)\, dy,$$
$$I(\zeta) = \prod_{i=1}^{s} I_i(\zeta) I_{s+i}(\zeta)$$

and

$$J(\mu) = \int_{E_n} I(\zeta) E(-\mu\zeta)\, dx,$$

where $\gamma \in \Gamma(t)$, $\gamma\alpha_i \to \mathfrak{a}_i$, $1 \le i \le 2s$, $\gamma_i = \alpha_i / a$ and $\gamma_{s+i} = \alpha_{s+i}/b$, $1 \le i \le s$. Then, by (9.8), the numbers γ_i satisfy

$$c_{10} < \gamma_i^{(\ell)} < c_{11}, \qquad c_{10} < |\gamma^{(m)}| < c_{11}, \qquad 1 \le i \le 2s. \tag{9.19}$$

Lemma 9.5. *We have*

$$\int_{B'_\gamma} S(\xi, T) E(-d\chi\xi)\, dx = G(\gamma) E(-d\chi\gamma) J(0) (T_1 T_2)^{sn} (ab)^{-n} M^{-20kny} + O\Big((T_1T_2)^{sn}(ab)^{-n} M^{-20kny-18y}\Big),$$

where

$$(9.20)\qquad \begin{cases} J(0) = D^{(1-2s)/2}k^{-2sn}N(\gamma_1,\dots,\gamma_{2s})^{-1/k}\prod_{\ell} F_\ell \prod_m H_m, \\ F_\ell = \int_{U_\ell} \prod_{i=1}^{2s} w_i^{\frac{1}{k}-1}\,dw, \end{cases}$$

in which $dw = \prod_{1\le i<2s} dw_i$ *and* U_ℓ *denotes the domain*

$$0 \le w_i \le \gamma_i^{(\ell)},\ (1\le i\le 2s), \qquad w_1+\dots+w_s-w_{s+1}-\dots-w_{2s}=0,$$

and where

$$H_m \int_{V_m} \prod_{i=1}^{2s} w_i^{\frac{1}{k}-1}\,dw\,d\varphi,$$

in which $d\varphi = \prod_{1\le i<2s} d\varphi_i$ *and* V_m *denotes the domain*

$$\begin{gathered} 0\le w_i \le |\gamma_i^{(\ell)}|^2\ (1\le i\le 2s), \quad -\pi\le\varphi_i\le\pi\ (1\le j\le 2s-1), \\ w_{2s} = |w_1^{\frac{1}{2}}e^{i\varphi_1}+\dots+w_{2s-1}^{\frac{1}{2}}e^{i\varphi_{2s-1}}|^2. \end{gathered}$$

For the proof of this lemma we require:

Lemma 9.6. *We have*

$$\int_{B'_\gamma} S(\xi,T)E(-d\chi\xi)\,dx = G(\gamma)E(-d\chi\gamma)\int_{E_n} I(\zeta,T)\,dx + O\Big((T_1T_2)^{sn}(ab)^{-n}M^{-20kny-18y}\Big).$$

Proof. By Lemmas 5.3 and 5.6, we have

$$\begin{aligned} I_j(\zeta,T_1) &\ll \prod_{i=1}^{n}\min\left(T_1, a^{-1/k}|\zeta^{(i)}|^{-1/k}\right), \\ I_{s+j}(\zeta,T_2) &\ll \prod_{i=1}^{n}\min\left(T_2, b^{-1/k}|\zeta^{(i)}|^{-1/k}\right), \end{aligned}$$

and

$$\begin{aligned} S_j(\xi,T_1) &= G_j(\gamma)I_j(\zeta,T_1) + O\big(M^{3y/2n}T_1^{n-1}\big), \\ S_{s+j}(\xi,T_2) &= G_{s+j}(\gamma)I_{s+j}(\zeta,T_2) + O\big(M^{3y/2n}T_2^{n-1}\big), \qquad 1\le j\le s. \end{aligned}$$

Therefore

$$S(\xi,T) = G(\gamma)I(\zeta,T) + O\Big((T_1T_2)^{sn}M^{3y/2n}\max(T_1^{-1}T_2^{-1})\Big).$$

By (5.4) with $\alpha=1$ we obtain

$$\int_{B'_\gamma} dx \ll (ht)^{-n} = (ab)^{-n} M^{-20kny+\frac{y}{2}}$$

and by (9.14)

$$\max(T_1^{-1}T_2^{-1} \ll M^{-20y}.$$

Therefore

$$\begin{aligned}(9.21)\quad \int_{B'_\gamma} S(\xi,T)E(-d\chi\xi)\,dx = G(\gamma)E(-d\chi\gamma)\int_{B'_\gamma} I(\zeta,T)E(-d\chi\zeta)\,dx \\ + O\Big((T_1T_2)^{sn}(ab)^{-n}M^{-20kny-18y}\Big),\end{aligned}$$

In the integral on the right hand side of (9.21) we replace $E(-d\chi\zeta)$ by 1. The error involved is

$$\begin{aligned}(T_1T_2)^{sn}\int_{B'_\gamma} \|\chi\zeta\|\,dx &\ll (T_1T_2)^{sn}H(ht)^{-n-1} \\ &\ll (T_1T_2)^{sn}(ab)^{-n}M^{-20kny-18y}.\end{aligned}$$

Hence

$$\begin{aligned}(9.22)\quad \int_{B'_\gamma} S(\xi,T)E(-d\chi\xi)\,dx = G(\gamma)E(-d\chi\gamma)\int_{B'_\gamma} I(\zeta,T)\,dx \\ + O\Big((T_1T_2)^{sn}(ab)^{-n}M^{-20kny-18y}\Big).\end{aligned}$$

If $\mathbf{x}$ is a point in $E_n - B'_\gamma$, then the inequality $h|\zeta^{(i)}| \geq t^{-1}$ is true for at least on index i. By (5.4),

$$\begin{aligned}&\int_{E_n-B'_\gamma} I(\zeta,T)\,dx \\ &\ll \int_{E_n-B'_\gamma}\left(\prod_{i=1}^{s}\min\left(T_1, a^{-1/k}|\zeta^{(i)}|^{-1/k}\right)\prod_{j=1}^{s}\min\left(T_2, b^{-1/k}|\zeta^{(i)}|^{-1/k}\right)\right)^{s} dx \\ &\ll \left(\int_{(ht)^{-1}}^{\infty}(ab)^{-s/k}u^{-2s/k}du\right)\times \\ &\quad\times\left(\int_0^\infty \min(T_1^s, a^{-s/k}v^{-s/k})\min(T_2^s, b^{-s/k}v^{-s/k})dv\right)^{r_1-1}\times \\ &\quad\times\left(\int_{-\pi}^{\pi}\int_0^\infty \min(T_1^{2s}, a^{-2s/k}w^{-2s/k})\min(T_2^{2s}, b^{-2s/k}w^{-2s/k})w\,dw\,d\varphi\right)^{r_2} \\ &\quad+\left(\int_0^\infty \min(T_1^s, a^{-s/k}u^{-s/k})\min(T_2^s, b^{-s/k}u^{-s/k})du\right)^{r_1}\times \\ &\quad\times\left(\int_{-\pi}^{\pi}\int_{(ht)^{-1}}^{\infty}(ab)^{-2s/k}v^{-4s/k+1}dv\,d\varphi\right) \\ &\quad\times\left(\int_{-\pi}^{\pi}\int_0^\infty \min(T_1^{2s}, a^{-2s/k}w^{-2s/k})\min(T_2^{2s}, b^{-2s/k}w^{-2s/k})w\,dw\,d\varphi\right)^{r_2-1}.\end{aligned}$$

Since, by (5.5) and (5.6),

$$\begin{aligned}\int_0^\infty \min(T_1^s, a^{-s/k}u^{-s/k}) \min(T_2^s, b^{-s/k}u^{-s/k})\, du \\ &\ll T_1^s \int_0^\infty \min(T_2^s, b^{-s/k}u^{-s/k})\, du \\ &\ll T_1^s T_2^{s-k} b^{-1}\end{aligned}$$

and

$$\int_0^\infty \min(T_1^{2s}, a^{-2s/k}w^{-2s/k}) \min(T_2^{2s}, b^{-2s/k}w^{-2s/k}) w\, dw \ll T_1^{2s} T_2^{2(s-k)} b^{-2},$$

we have, for $s > q\,(> 20kn)$,

$$\begin{aligned}&\int_{E_n - B'_\gamma} I(\zeta, T)\, dx \\ &\ll (ht)^{\frac{2s}{k}-1}(ab)^{-s/k} T_1^{s(r_1-1)} b^{-(r_1-1)} T_2^{(s-k)(r_1-1)} T_1^{2sr_2} b^{-2r_2} T_2^{2(s-k)r_2} \\ &\quad + T_1^{sr_1} b^{-r_1} T_2^{(s-k)r_1} (ht)^{\frac{4s}{k}-2} (ab)^{-2s/k} T_1^{2s(r_2-1)} b^{-2(r_2-1)} T_2^{2(s-k)(r_2-1)} \\ &\ll (ht)^{\frac{2s}{k}-1}(ab)^{-s/k} b^{-n+1} T_1^{s(n-1)} T_2^{(s-k)(n-1)} \\ &\quad + (ht)^{\frac{4s}{k}-2}(ab)^{-2s/k} b^{-n+2} T_1^{s(n-2)} T_2^{(s-k)(n-2)} \\ &\ll (T_1T_2)^{sn}(ab)^{-n} M^{-20kny} \left(M^{\frac{y}{2n}-\frac{sy}{kn}} + M^{\frac{y}{n}-\frac{2sy}{kn}}\right) \\ &\ll (T_1T_2)^{sn}(ab)^{-n} M^{-20kny-18y}.\end{aligned}$$

The lemma follows by substituting this into (9.21). □

Proof of Lemma 9.5. Let

$$\begin{aligned} \eta' &= y_1w_1 + \cdots + y'_n w_n, & \zeta' &= x_1\rho_1 + \cdots + x'_n\rho_n, \\ dy' &= dy'_1 \cdots dy'_n, & dx' &= dx'_1 \cdots dx'_n, \\ \eta &= T_1\eta' \text{ in } I_i(\zeta, T_1), & \eta &= T_2\eta' \text{ in } I_{s+i}(\zeta, T_2), \quad 1 \le i \le s, \end{aligned}$$

and $\zeta = a^{-1}b^{-1}M^{-20ky}\zeta'$. Then

$$\alpha_i \eta^k \zeta = \gamma_i {\eta'}^k \zeta', \qquad 1 \le i \le 2s.$$

Writing η and ζ for η' and ζ' again, we find that

$$\int_{E_n} I(\zeta, T)\, dx = (T_1T_2)^{sn}(ab)^{-n} M^{-20kny} J(0),$$

and the lemma follows by Theorem 6.2 and Lemma 9.6. □

9.8 Proof of Theorem 9.1

We have
$$\sum_{\gamma\in\Gamma(t)} 1 \ll \sum_{N(\mathfrak{a})\le t^n} N(\mathfrak{a}) \ll \sum_{d\le t^n} d^2 \ll t^{3n} = M^{3y},$$
and, by Lemma 3.2,
$$H^n \ll \sum_{\chi\in P(H)} 1 \ll H^n. \tag{9.22}$$

Therefore, by (9.6), (9.11), (9.15) and Lemmas 9.4 and 9.5, we have
$$\begin{aligned} R &= \sum_{\gamma\in\Gamma(t)} \int_{B'_\gamma} F(\xi)\,dx + O\big(H^n(T_1T_2)^{sn}(ab)^{-n}M^{-20kny-15y}\big) \\ &= J(0)\mathfrak{S}(t,H)(T_1T_2)^{sn}(ab)^{-n}M^{-20kny} \\ &\qquad\qquad + O\big(H^n(T_1T_2)^{sn}(ab)^{-n}M^{-20kny-15y}\big), \end{aligned}$$
where $J(0)$ is defined by (9.20) and
$$\mathfrak{S} = \mathfrak{S}(t,H) = \sum_{\chi\in P(H)} \sum_{\gamma\in\Gamma(t)} G(\gamma)E(-d\chi\gamma).$$

Let $\Sigma^\star_\gamma$ denote a sum with γ running over a reduced residue system $(\mathfrak{a}\delta)^{-1}$ mod δ^{-1}. Then
$$\begin{aligned} \mathfrak{S} &= \sum_{N(\mathfrak{a})=1} \sum_{\gamma}{}^{\star} G(\gamma) \sum_{\chi\in P(H)} E(-d\chi\gamma) + \sum_{1\le N(\mathfrak{a})\le t^n} \sum_{\gamma}{}^{\star} G(\gamma) \sum_{\chi\in P(H)} E(-d\chi\gamma) \\ &= \mathfrak{S}_1 + \mathfrak{S}_2, \end{aligned}$$
say. Now $\mathfrak{S}_1 \gg H^n$ by (9.22). If $N(\mathfrak{a}) > 1$, then, by Lemma 2.5,
$$\sum_{\chi(\mathfrak{a})} E(-d\chi\gamma) = 0.$$

For any given integer μ, it follows from the definition of t, H and Lemma 3.2 that the number of $\nu \in \mathfrak{a}$ and $\nu + \mu \in P(H)$ is
$$\frac{(2\pi)^{r_2}}{\sqrt{D}\,N(\mathfrak{a})} H^n + O\Big(\frac{H^{n-1}}{N(\mathfrak{a})^{1-1/n}}\Big).$$

Hence, if the domain $\chi \in P(H)$ has to be split up into a union of complete residue sets mod $\mathfrak{a}$ plus a few other remaining, say W, elements, then
$$W \ll N(\mathfrak{a}) \frac{H^{n-1}}{N(\mathfrak{a})^{1-1/n}} \ll N(\mathfrak{a})^{\frac{1}{n}} H^{n-1},$$

and thus

$$\begin{aligned}\mathfrak{S}_2 &\ll \sum_{N(\mathfrak{a})\le t^n} \sum_{\gamma}{}^{\star} W \ll H^{n-1} \sum_{N(\mathfrak{a})\le t^n} \sum_{\gamma}{}^{\star} N(\mathfrak{a})^{\frac{1}{n}} \\ &\ll H^{n-1} \sum_{N(\mathfrak{a})\le t^n} N(\mathfrak{a})^{1+\frac{1}{n}} \ll H^{n-1} \sum_{d\le t^n} d^3 \ll H^{n-1} t^{4n} \ll H^n M^{-2y}.\end{aligned}$$

Consequently we have $\mathfrak{S} \gg H^n$. Since $J(0) \gg c_{14}(k, \mathbb{K}, q, y)$ by (9.19), we have

$$R \gg c_{15}(k, \mathbb{K}, q, y) H^n (T_1 T_2)^{sn} (ab)^{-n} M^{-20kny} > 1,$$

if $M \ge c_8(k, \mathbb{K}, x', \epsilon)$. The theorem is proved. □

Notes

Theorem 9.1 is a generalisation of a theorem of Schmidt; see Wang [1]. Schmidt [1] first proved that if $a_i, b_i\,(1 \le i \le s)$ are positive integers and $s \ge c_{16}(k, \epsilon)$, then the equation

$$a_1 x_1^k + \cdots + a_s x_s^k - b_1 y_1^k - \cdots - b_s y_s^k = 0$$

has nonnegative integers solution x_i, y_i not all zero and satisfying

$$\max_{i,j}(x_i, y_j) \le \max_{i,j}(a_i, b_j)^{\frac{1}{k}+\epsilon}.$$

This is an improvement on a result of Birch [4]. Let $a_1 = \cdots = a_s = a$ and $b_1 = \cdots = b_s = b$ with $(a, b) = 1$. Then

$$\max_{i,j}(x_i, y_j) \ge s^{-\frac{1}{k}} \max(a, b)^{\frac{1}{k}},$$

so that Schmidt's theorem is best possible apart from the improvement on the factor $\max_{i,j}(a_i, b_j)^{\epsilon}$.

Lemma 9.1: See Hurwitz [1].

Chapter 10

Small Solutions of Additive Equations

10.1 Introduction

We assume that $\mathbb{K}$ is a purely imaginary algebraic number field of degree $n = 2r_2$ in these last two chapters. We write r for r_2, and we suppose that $\mathbb{K}^{(m)}$ and $\mathbb{K}^{(m+r)}$ are complex conjugate of $\mathbb{K}$.

Let $\alpha_1, \ldots, \alpha_s$ be a set of integers in $\mathbb{K}$. Consider the additive form

$$A(\boldsymbol{\lambda}) = \sum_{i=1}^{n} \alpha_i \lambda_i^k. \tag{10.1}$$

Write

$$A = \max(1, \|\boldsymbol{\alpha}\|).$$

Note that the definition of A is distinct from that given in §5.1.

Theorem 10.1. *Let $s \geq c_1(k, r, \epsilon)$. Then the equation*

$$A(\boldsymbol{\lambda}) = 0 \tag{10.2}$$

has a solution $\boldsymbol{\lambda} \neq \mathbf{0}$, $\boldsymbol{\lambda} \in J^s$, satisfying

$$\|\boldsymbol{\lambda}\| \ll A^{\epsilon}. \tag{10.3}$$

10.2 Reductions

If $A(\boldsymbol{\lambda})$ is a form not identically zero, then we put

$$A'(\boldsymbol{\lambda}) = \frac{c_2}{\sigma} A(\boldsymbol{\lambda}),$$

where σ is a nonzero element in the integral ideal $(\alpha_1, \ldots, \alpha_s)$ with least norm in absolute value, and $c_2(\mathbb{K})$ is the constant $c_4(\mathbb{K})$ as stated in Lemma 9.2. Then $\sigma | c_2 \alpha_i$, $1 \leq i \leq s$. We may assume without loss of generality that

$$|N(\sigma)|^{\frac{1}{2r}} \ll \|\sigma\| \ll |N(\sigma)|^{\frac{1}{2r}},$$

because we can choose a unit η such that $\eta\sigma$ satisfies the above relation, and use $\eta\sigma$ instead of σ; see Lemma 3.1. If $A(\boldsymbol{\lambda}) \equiv 0$, then we may put $A'(\boldsymbol{\lambda}) = A(\boldsymbol{\lambda})$.

It follows from Theorem 8.3 that if $s \geq c_3(k,r)$, then (10.2) has a nontrivial solution satisfying

$$\|\boldsymbol{\lambda}\| \ll A^{c_4(k,r)}. \tag{10.4}$$

Let X be the set of x with the property that if $s \geq c_5(k,r,x)$, then (10.2) has a nontrivial solution satisfying $\|\boldsymbol{\lambda}\| \ll A^x$. From (10.4) we see that X is non-empty, and we denote its infimum by x. The conclusion of the theorem is that $x = 0$, and we now give an indirect proof by supposing that $x > 0$.

The polynomial

$$f(w) = x + kx^2 - kxw - k^2x^2w - w$$

satisfies $f(x) = -k^2x^3 < 0$. Therefore we may choose y such that

$$0 < y < x \qquad \text{and} \qquad x + kx^2 - kxy - k^2x^2y < y. \tag{10.5}$$

Take z so small that

$$y + 12xz < x, \qquad z < y/10, \qquad z < 1/10. \tag{10.6}$$

Then pick x' satisfying

$$\max\left(y + 12xz,\, x - \frac{xz}{6r}\right) < x' < x. \tag{10.7}$$

We proceed to prove that $x' \in X$. It suffices to prove this for large A, say $A > c_6(k, \mathbb{K}, x')$. In fact, if $A \leq c_6$ and $s \geq c_3$, then it follows from (10.4) that (10.2) has a nontrivial solution such that

$$\|\boldsymbol{\lambda}\| \ll A^{c_4(k,r)} \ll c_6^{c_4} \ll A^{x'}.$$

Clearly we may suppose that $\alpha_i \neq 0$, $1 \leq i \leq s$. Finally, pick x'' satisfying

$$\max\left(y + 12xz,\, x - \frac{xz}{6r}\right) < x'' < x' \tag{10.7}$$

and then choose ϵ_1 such that

$$(1 + \epsilon_1)x'' + \frac{4\epsilon_1}{k} < x'. \tag{10.9}$$

By (5.9) we have

$$\min_{i,m} |\alpha_i^{(m)}| \geq A^{-2r+1}.$$

Divide the interval $[-2r+1, 1]$ into a finite family of intervals I with lengths at most ϵ_1. One of these intervals, say I_1, will have at least s_1 numbers among α_i satisfying

$$|\alpha_i^{(1)}| = A^{a_1}, \qquad a_1 \in I_1,$$

where $s_1 \geq s/([\frac{2r}{\epsilon_1}]+1)$. We may suppose without loss of generality that $\alpha_i, \ldots, \alpha_{s_1}$ satisfy the above relation. Similarly there is an interval I_2 in the family with at least s_2 numbers in $\alpha_1, \ldots, \alpha_{s_1}$ satisfying

$$|\alpha_i^{(2)}| = A^{a_2}, \qquad a_2 \in I_2,$$

where $s_1 \geq s_1/([\frac{2r}{\epsilon_1}]+1) \geq s/([\frac{2r}{\epsilon_1}]+1)^2$. We may suppose that $\alpha_1, \ldots, \alpha_{s_2}$ have the above property. Continuing this process, we obtain t numbers among $\alpha_1, \ldots, \alpha_s$, which we may suppose to be $\alpha_1, \ldots, \alpha_t$, satisfying

$$\left|\frac{\alpha_i^{(m)}}{\alpha_j^{(m)}}\right| \leq A^{\epsilon_1}, \qquad 1 \leq i,j \leq t, \tag{10.10}$$

where $t \geq s/([\frac{2r}{\epsilon_1}]+1)^r$.

Suppose that, if $\alpha_1, \ldots, \alpha_t$ satisfy (10.10) and $t \geq c_7(k,r,x')$, the equation

$$\alpha_1 \lambda_1'^k + \cdots + \alpha_t \lambda_t'^k = 0$$

has a nontrivial solution satisfying

$$\max_i \|\lambda_i'\| \ll A^{x'}.$$

Take $c_5(k,r,x') = ([\frac{2r}{\epsilon_1}]+1)^r c_7(k,r,x')$ and set $\lambda_i = \lambda_i' \, (1 \leq i \leq t)$ and $\lambda_j = 0 \, (t < j \leq s)$. We have a nontrivial solution of (10.2) with

$$\|\boldsymbol{\lambda}\| \ll A^{x'}$$

if $s \geq c_5$. Hence we may suppose that the coefficients in equation (10.2) satisfy

$$\left|\frac{\alpha_i^{(m)}}{\alpha_j^{(m)}}\right| \leq A^{\epsilon_1}, \qquad 1 \leq i,j \leq s. \tag{10.11}$$

10.3 Continuation

By Lemma 3.1, there exists a set of units $\sigma_i \, (1 \leq i \leq s)$ such that

$$|N(\alpha_i)|^{\frac{1}{2r}} \ll |\alpha_i^{(m)} \sigma_i^{(m)k}| \ll |N(\alpha_i)|^{\frac{1}{2r}}, \qquad 1 \leq i \leq s. \tag{10.12}$$

Let $a^{2r} = A^{2r\epsilon_1} \max |N(\alpha_i)|$ and let p_i be the largest rational integer satisfying

$$|N(\alpha_i)| p_i^{2kr} \leq a^{2r}, \qquad 1 \leq i \leq s.$$

Since $A \geq c_6$ and $a^{2r}/|N(\alpha_i)| \geq A^{2r\epsilon_1}$, we may suppose that

$$p_i \geq 2^{-\frac{1}{2kr}} \left(\frac{a^{2r}}{|N(\alpha_i)|} \right)^{\frac{1}{2kr}}, \qquad 1 \leq i \leq s.$$

Hence

$$|N(\alpha_i)| p_i^{2kr} \geq \tfrac{1}{2} a^{2r}, \qquad 1 \leq i \leq s.$$

Set

$$\alpha_i' = \alpha_i \sigma_i^k p_i^k \qquad \lambda_i = \sigma_1^{-1} \sigma_i p_i \lambda_i', \qquad 1 \leq i \leq s. \tag{10.13}$$

Then

$$A_1(\boldsymbol{\lambda}') = \alpha_1' \lambda_1'^k + \cdots + \alpha_s' \lambda_s'^k = \sigma_1^k A(\boldsymbol{\lambda}),$$

where $\boldsymbol{\lambda}' = (\lambda_1', \ldots, \lambda_s')$. By (10.11) and (10.12), we have

$$a \ll |\alpha_i'^{(m)}| \ll a, \qquad 1 \leq i \leq s,$$

and

$$\begin{aligned} |\sigma_1^{(m)-1} \sigma_i^{(m)}| &= \left| \frac{\sigma_i^{(m)k} \alpha_i^{(m)} \alpha_i^{(m)-1}}{\sigma_1^{(m)k} \alpha_1^{(m)} \alpha_1^{(m)-1}} \right|^{\frac{1}{k}} \\ &\ll |N(\alpha_i)|^{1/2kr} |N(\alpha_1)|^{-1/2kr} A^{\epsilon_1/k} \\ &\ll A^{2\epsilon_1/k}, \qquad 1 \leq i \leq s. \end{aligned} \tag{10.14}$$

Since

$$\begin{aligned} |N(\alpha_i)| &= A^{2r\epsilon_1} |N(\alpha_i)| \max_j |N(\alpha_j)| \Big/ A^{2r\epsilon_1} |N(\alpha_i)| \max_q |N(\alpha_q)| \\ &= a^{2r} |N(\alpha_i)| A^{-2r\epsilon_1} \Big/ \max_q |N(\alpha_q)| \geq a^{2r} A^{-4r\epsilon_1}, \end{aligned}$$

we have

$$p_i^{2kr} \leq \frac{a^{2r}}{|N(\alpha_i)|} \leq A^{4r\epsilon_1}, \qquad 1 \leq i \leq s. \tag{10.15}$$

Suppose that the equation $A_1(\boldsymbol{\lambda}') = 0$ has a nontrivial solution satisfying

$$\|\boldsymbol{\lambda}'\| \ll \max_i \|\alpha_i'\|^{x''} \ll a^{x''} \ll A^{(1+\epsilon_1)x''}.$$

Then it follows by (10.9), (10.13), (10.14) and (10.15) that (10.2) has a nontrivial solution satisfying

$$\|\boldsymbol{\lambda}'\| \ll A^{(1+\epsilon_1)x'' + 4\epsilon_1/k} \ll A^{x'}.$$

Thus it suffices to prove that if $s \geq c_8$ and α_i satisfy

$$c_9 a < |\alpha_i^{(m)}| < c_{10} a, \qquad 1 \leq i \leq s, \tag{10.16}$$

where $c_9 = c_9(k, \mathbb{K})$, $c_{10} = c_{10}(k, \mathbb{K})$ and $a > c_6(k, \mathbb{K}, x'')$, then (10.2) has a nontrivial solution with

$$\|\boldsymbol{\lambda}\| \ll a^{x''}. \tag{10.17}$$

Although the constant c_8 depends on k, r and x'' we shall not indicate the dependency of such constants in what follows because the parameters k, r, x, y, z, x', x'' are considered to be fixed.

We first show that the above assertion can be derived from the following:

Proposition 1. *Let $\alpha_i\,(1 \le i \le s)$ satisfy (10.6). If $s \ge c_{11}$, then either (10.2) has a nontrivial solution satisfying (10.17), or there is a nonzero integer χ such that*

$$A(\boldsymbol{\lambda}) = \chi, \qquad \chi \in P(a^{6z}), \qquad \lambda_i \in P(a^y), \qquad 1 \le i \le s, \tag{10.18}$$

where $\boldsymbol{\lambda} \in J^s$.

We may suppose that $c_5(k, r, 2x)$ and c_{11} are integers. Denote by $v = c_5$, $u = c_{11}$ and $s = uv$. Replace the indices $1 \le i \le s$ by double indices $1 \le i \le v$, $1 \le j \le u$. Then (10.1) becomes

$$A(\boldsymbol{\lambda}) = \sum_{i=1}^{v} A_i(\boldsymbol{\lambda}_i),$$

where $\boldsymbol{\lambda}_i = (\lambda_{i1}, \ldots, \lambda_{iu})$ and

$$A_i(\boldsymbol{\lambda}_i) = \sum_{j=1}^{u} \alpha_{ij}\lambda_{ij}^k, \qquad 1 \le i \le v.$$

If there is an equation, say $A_i(\boldsymbol{\lambda}_i) = 0$, which has a nontrivial solution having

$$\|\boldsymbol{\lambda}_i\| \ll a^{x''},$$

then we have directly a nontrivial solution of (10.2) satisfying (10.17). Otherwise it follows by Proposition 1 that there are nonzero integers $\chi_1, \ldots, \chi_v$ satisfying

$$A_i(\boldsymbol{\lambda}_i) = \chi_i, \qquad \chi_i \in P(a^{6z}), \qquad \lambda_{ij} \in P(a^y), \qquad 1 \le i \le v,\ 1 \le j \le u,$$

where λ_{ij} are integers. Since the equation

$$B(\boldsymbol{\mu}) = \chi_1\mu_1^k + \cdots + \chi_v\mu_v^k = 0$$

has a nontrivial solution with

$$\|\boldsymbol{\mu}\| \ll \max \|\chi_i\|^{2x} \ll a^{12xz},$$

equation (10.2) has a nontrivial solution satisfying

$$\max_{i,j} \|\mu_i \lambda_{ij}\| = \|\boldsymbol{\mu}\| \max_i \|\boldsymbol{\lambda}_i\| \ll a^{y+12xz} \ll a^{x''}$$

by (10.8). It remains therefore to prove Proposition 1.

10.4 Farey Divisions

Let

$$h = a^{1+ky-\frac{z}{2r}} \qquad \text{and} \qquad t = a^{\frac{z}{3r}}.$$

We take the Farey division of U_n with respect to (h, t) and define $\Gamma(t), B_\gamma, B$ and S; see §5.1. Let B'_γ denote the set

$$\{\mathbf{x} : \mathbf{x} \in U_n,\ h\|\xi - \gamma_0\| \le t^{-1} \le N(\mathfrak{a})^{-1/n} \quad \text{for some} \quad \gamma_0 \equiv \gamma \ (\bmod\ \delta^{-1})\}.$$

Then $B'_\gamma \subset B_\gamma\, (\gamma \in \Gamma(t))$, and it follows from Lemma 5.1 that if $\gamma_1, \gamma_2 \in \Gamma(t)$ and $\gamma_1 \ne \gamma_2$, then $B'_{\gamma_1} \cap B'_{\gamma_2} = \phi$. Set

$$B' = \bigcup_{\gamma \in \Gamma(t)} B'_\gamma$$

and define the supplementary domain S' of B' with respect to U_n by

$$S' = U_n - B'.$$

Then $S' \supset S$. Further, let

$$T = a^y,\ H = a^{6z} \quad \text{and} \quad F(\xi) = \sum_{\chi \in P(H)} S(\xi, T) E(-\chi\xi), \tag{10.19}$$

where $S(\xi, T)$ is defined in §5.1. Denote by R the number of solutions of the equation

$$A(\boldsymbol{\lambda}) = \chi$$

in integers $\lambda_1, \ldots, \lambda_s, \chi$ satisfying

$$\lambda_i \in P(T)\ (1 \le i \le s), \qquad \chi \in P(H).$$

Then

$$R = \int_{B'} F(\xi)\, dx + \int_{S'} F(\xi)\, dx. \tag{10.20}$$

We shall show that, under the assumption made in Proposition 1, either (10.2) has a nontrivial solution satisfying (10.17), or else $R > 1$.

10.5 The Supplementary Domain

Lemma 10.1. *Suppose that* $s \geq c_{12}$ *and* $\mathbf{x} \in S'$. *Then either*

$$|F(\xi)| < H^{2r}T^{2(s-k)r}a^{-4r} \tag{10.21}$$

or there is a nontrivial solution of (10.2) *satisfying* (10.17).

Proof. Take ϵ_2 such that

$$0 < \epsilon_2 < c_{13} < \frac{1}{2G}, \tag{10.22}$$

where $G = 2^{k-1}$ and c_{13} is a constant to be specified later. Set

$$u = c_5(k, r, x + \epsilon_2) \qquad \text{and} \qquad q = u^2. \tag{10.23}$$

Choose $c_{12}\,(> 8kn)$ sufficiently large so that, if $s \geq c_{12}$, then

$$\frac{2r(k+2y^{-1})}{s-q+1} < \epsilon_2;$$

we now have, by (10.19),

$$(T^k a^2)^{\frac{2r}{s-q+1}} = T^{\frac{2r(k+2y^{-1})}{s-q+1}} < T^{\epsilon_2}.$$

If (10.21) fails to hold, then

$$H^{2r}|S_1(\xi,T)\cdots S_s(\xi,T)| \gg |F(\xi)| \geq H^{2r}T^{2(s-k)r}a^{-4r}.$$

We may suppose without loss of generality that

$$|S_1(\xi,T)| \geq \cdots \geq |S_s(\xi,T)|.$$

Hence we have

$$|S_q(\xi,T)|^{s-q+1}T^{2(q-1)r} \gg T^{2(s-k)r}a^{-4r},$$

and thus, by (10.22),

$$\begin{aligned} |S_i(\xi,T)| &\gg T^{2(s-q-k+1)r/(s-q+1)}a^{-4r/(s-q+1)} \\ &\gg T^{2r}(T^k a^2)^{-2r/(s-q+1)} > T^{2r-\epsilon_2} \\ &> T^{2r-\frac{1}{G}+\epsilon_2}, \qquad 1 \leq i \leq q. \end{aligned}$$

On taking $C = T^{2r-\epsilon_2}$ it follows from Theorem 3.3 that there are integers $\sigma_i, \beta_i\,(1 \leq i \leq s)$ such that

$$\|\xi\alpha_i\sigma_i - \beta_i\| \ll T^{-k+2G\epsilon_2}, \quad 0 < \|\sigma_i\| \ll T^{2G\epsilon_2}, \qquad 1 \leq i \leq q. \tag{10.24}$$

Denote by $\tau_i = \beta_i\sigma_i^{k-1}$, $1 \leq i \leq q$. Then

$$\|\xi\alpha_i\sigma_i^k - \tau_i\| \le \|\sigma_i\|^{k-1}\|\xi\alpha_i\sigma_i - \beta_i\| \ll T^{-k+2kG\epsilon_2}, \qquad 1 \le i \le q.$$

Therefore, by (10.16), we have

$$\begin{aligned}\|\alpha_i\sigma_i^k\tau_j - \alpha_j\sigma_j^k\tau_i\| &\le \|(\xi\alpha_j\sigma_j^k - \tau_j)\alpha_i\sigma_i^k\| + \|(\xi\alpha_i\sigma_i^k - \tau_i)\alpha_j\sigma_j^k\| \\ &\ll aT^{-k+4kG\epsilon_2}, \qquad 1 \le i,j \le q;\end{aligned}$$

in other words, the vector $\mathbf{f}_i = (\alpha_i\sigma_i^k, \tau_i)$, $1 \le i \le q$, satisfy

$$\|\det(\mathbf{f}_i, \mathbf{f}_j)\| \ll aT^{-k+4kG\epsilon_2}, \qquad 1 \le i,j \le q. \tag{10.25}$$

Let β be a nonzero element in the integral ideal $(\alpha_1\sigma_1^k, \tau_1)$ with the least norm in absolute value. Then $\beta | c_2\alpha_1\sigma_1^k$ and $\beta | c_2\tau_1$ by Lemma 9.2. Set

$$c_2\alpha_1\sigma_1^k = \beta\sigma, \quad c_2\tau_1 = \beta\tau \quad \text{and} \quad \mathbf{f} = (\sigma, \tau).$$

Then

$$c_2\mathbf{f}_1 = \beta\mathbf{f}.$$

By Lemma 3.1, we may choose β such that

$$|N(\sigma)|^{\frac{1}{2r}} \ll \|\sigma\| \ll |N(\sigma)|^{\frac{1}{2r}}. \tag{10.26}$$

We also have two integers σ' and τ' such that

$$\beta = \alpha_1\sigma_1^k\tau' - \tau_1\sigma';$$

therefore $c_2\beta = \beta\sigma\tau' - \beta\tau\sigma'$, that is

$$c_2 = \sigma\tau' - \tau\sigma'. \tag{10.27}$$

Set $\mathbf{g} = (\sigma', \tau')$. Then, by (10.27),

$$c_2\mathbf{f}_i = \varphi_i\mathbf{f} + \psi_i\mathbf{g},$$

where

$$\varphi_i = \begin{vmatrix} \alpha_i\sigma_i^k & \sigma' \\ \tau_i & \tau' \end{vmatrix} \quad \text{and} \quad \psi_i = \begin{vmatrix} \sigma & \alpha_i\sigma_i^k \\ \tau & \tau_i \end{vmatrix} \qquad (1 \le i \le q)$$

are integers. By (5.9) and (10.24), we have

$$|\sigma_i^{(m)}| \ge 2^{-2(2r-1)G\epsilon_2}, \qquad 1 \le i \le q. \tag{10.28}$$

Since $c_2\alpha_1\sigma_1^k = \beta\sigma$, we have, by (10.16), (10.24), (10.26) and (10.28) that

$$a\|\sigma\|^{-1}T^{2kG\epsilon_2} \gg |\beta^{(m)}| \gg a\|\sigma\|^{-1}T^{-2k(2r-1)G\epsilon_2}. \tag{10.29}$$

Therefore it follows from (10.25) that

$$\begin{aligned}\|\psi_i\| = \|\det(\mathbf{f}_i, \mathbf{f})\| &\le c_2 \max_m |\beta^{(m)}|^{-1}\|\det(\mathbf{f}_i, \mathbf{f}_1)\| \\ &\le \|\sigma\|T^{-k+8krG\epsilon_2} = M \text{ (say)} \qquad 1 \le i \le q.\end{aligned} \tag{10.30}$$

1) Suppose that $M \geq 1$. Replace the indices $1 \leq i \leq q$ by double indices $1 \leq i, j \leq u$. Define

$$A_i(\boldsymbol{\lambda}_i) = \sum_{j=1}^{u} \psi_{ij} \lambda_{ij}^k, \qquad 1 \leq i \leq u,$$

where $\boldsymbol{\lambda}_i = (\lambda_{i1}, \ldots, \lambda_{iu})$, $1 \leq i \leq u$. It follows from (10.23), (10.30) and the definition of the set X that the equation $A_i(\boldsymbol{\lambda}_i) = 0$ has a nontrivial solution satisfying

$$\|\boldsymbol{\lambda}_i\| \ll M^{x+\epsilon_2}, \qquad 1 \leq i \leq u. \tag{10.31}$$

Let

$$\mathbf{g}_i = \sum_{j=1}^{u} \lambda_{ij}^k \mathbf{f}_{ij} = c_2^{-1} \Big(\sum_{j=1}^{u} \lambda_{ij}^k \varphi_{ij} \Big) \mathbf{f}, \qquad 1 \leq i, \leq u.$$

The first coordinate of $\mathbf{g}_i$ $(1 \leq i \leq u)$ is

$$\beta_i = c_2^{-1} \sigma \sum_{j=1}^{u} \lambda_{ij}^k \varphi_{ij}.$$

Therefore $c_2 \beta_i / \sigma$ $(1 \leq i \leq u)$ are integers. If $\beta_1, \ldots, \beta_u$ are not all zero, then let χ be a nonzero element in the integral ideal $(\beta_1, \ldots, \beta_u)$ with the least norm in absolute value satisfying $|N(\chi)|^{1/2r} \ll \|\chi\| \ll |N(\chi)|^{1/2r}$. Then $\sigma | c_2 \chi$ and, by (10.26), we have $\|\sigma\| \ll \|\chi\|$. Consider the form

$$B(\boldsymbol{\mu}) = \sum_{i=1}^{u} \beta_i \mu_i^k,$$

where $\boldsymbol{\mu} = (\mu_1, \ldots, \mu_u)$. Since

$$\beta_i = \sum_{j=1}^{u} \lambda_{ij}^k \alpha_{ij} \sigma_{ij}^k, \qquad 1 \leq i \leq u,$$

we have, by (10.16), (10.24) and (10.31),

$$\|\beta_i\| \ll a T^{2kG\epsilon_2} M^{(x+\epsilon_2)k}, \qquad 1 \leq i \leq u,$$

and therefore the equation $B(\boldsymbol{\mu}) = 0$ has a nontrivial solution satisfying

$$\|\boldsymbol{\mu}\| \ll \max \left(1, \frac{a T^{2kG\epsilon_2} M^{(x+\epsilon_2)k}}{\|\sigma\|} \right)^{x+\epsilon_2}.$$

Consequently (10.2) has a nontrivial solution

$$\mu_i \sigma_{ij} \lambda_{ij} \ (1 \leq i, j \leq u), \qquad \lambda_i = 0 \ (q < i \leq s)$$

satisfying

$$\begin{aligned}\|\boldsymbol{\lambda}\| &\ll T^{2G\epsilon_2} M^{x+\epsilon_2} \max\left(1, \frac{aT^{2kG\epsilon_2}M^{(x+\epsilon_2)k}}{\|\sigma\|}\right)^{x+\epsilon_2} \\ &\ll T^{2G\epsilon_2} \max\left(M, \frac{aT^{2kG\epsilon_2}M^{(x+\epsilon_2)k+1}}{\|\sigma\|}\right)^{x+\epsilon_2} \\ &= T^{2G\epsilon_2} I^{x+\epsilon_2}, \quad \text{say.}\end{aligned}$$

Since $c_2\alpha_1\sigma_1^k = \beta\sigma$, by (10.24), (10.26) and (10.30), we have

$$\begin{aligned}&\|\sigma\| \ll aT^{2kG\epsilon_2} \min|\beta^{(m)}|^{-1} \ll aT^{2kG\epsilon_2}, \\ &M \ll aT^{-k+10krG\epsilon_2},\end{aligned}$$

and therefore

$$\begin{aligned}I &\ll aT^{2kG\epsilon_2}M^{(x+\epsilon_2)k+1}\|\sigma\|^{-1} \\ &= aT^{2kG\epsilon_2}M^{(x+\epsilon_2)k+1}M^{-1}T^{-k+8krG\epsilon_2} \\ &\ll aT^{-k+10krG\epsilon_2}(aT^{-k+10krG\epsilon_2})^{(x+\epsilon_2)k}.\end{aligned}$$

Since $a > c_6$, we have

$$\|\boldsymbol{\lambda}\| \ll a^{x+kx^2-kxy-k^2x^2y+c_{14}(k,r)\epsilon_2}.$$

Hence, if c_{13} is sufficiently small, then, by (10.5) and (10.8), we have the desired result (10.17).

2) Suppose that $M < 1$. We revert to the single indices $1 \le i \le q$. By (10.30), we have $\psi_i = 0\,(1 \le i \le q)$; in other words $c_2\mathbf{f}_i$ are multiples of the integral vector $\mathbf{f}$. Let

$$B(\boldsymbol{\mu}) = \sum_{i=1}^{q} \alpha_i\sigma_i^k\mu_i^k,$$

where $\boldsymbol{\mu} = (\mu_1, \ldots, \mu_q)$. Let χ be a nonzero element in the integral ideal $(\alpha_1\sigma_1^k, \ldots, \alpha_q\sigma_q^k)$ with the least norm in absolute value satisfying $|N(\chi)|^{1/2r} \gg \|\chi\| \gg |N(\chi)|^{1/2r}$. Then $\sigma|c_2\chi$, and so $\|\sigma\| \ll \|\chi\|$. Hence the equation $B(\boldsymbol{\mu}) = 0$ has a nontrivial solution satisfying

$$\|\boldsymbol{\mu}\| \ll \max_i \left(\|\alpha_i\sigma_i^k\|\,\|\sigma\|^{-1}\right)^{x+\epsilon_2} \ll \left(aT^{2kG\epsilon_2}\|\sigma\|^{-1}\right)^{x+\epsilon_2}.$$

This gives a nontrivial solution of (10.2):

$$\lambda_i = \sigma_i\mu_i \ (1 \le i \le q), \qquad \lambda_j = 0 \ (q < j \le s).$$

If c_{13} is sufficiently small, then

$$\begin{aligned}\|\boldsymbol{\lambda}\| &\ll T^{2G\epsilon_2}\left(aT^{2kG\epsilon_2}\|\sigma\|^{-1}\right)^{x+\epsilon_2} \\ &= a^{x+\epsilon_2+2Gy\epsilon_2+2kGy(x+\epsilon_2)\epsilon_2}\|\sigma\|^{-x-\epsilon_2} \\ &\ll a^{x+\frac{xz}{6r}}\|\sigma\|^{-x}.\end{aligned}$$

If $\|\sigma\| \geq a^{z/3r}$, then, by (10.8), we have

$$\|\boldsymbol{\lambda}\| \ll a^{x-\frac{xz}{6r}} \ll a^{x''},$$

so that (10.17) holds. Now suppose that $\|\sigma\| < a^{z/3r}$. Then, by (10.24) and (10.28), we have

$$\begin{aligned}\|\xi - \sigma^{-1}\tau\| &= \|\xi - \alpha_1^{-1}\sigma_1^{-1}\beta_1\| \leq \max_m |\alpha_1^{(m)}\sigma_1^{(m)}|^{-1}\|\alpha_1\sigma_1\xi - \beta_1\| \\ &\ll a^{-1}T^{2(2r-1)G\epsilon_2 - k + 2G\epsilon_2} < a^{-1-ky+\frac{z}{6r}} = h^{-1}t^{-1},\end{aligned}$$

if c_{13} is sufficiently small. Let $\sigma^{-1}\tau\delta = \mathfrak{b}/\mathfrak{a}$, $(\mathfrak{a}, \mathfrak{b}) = 1$. Then $\mathfrak{a}|\sigma$ and thus

$$N(\mathfrak{a}) \leq |N(\sigma)| < a^{2z/3} = t^{2r}.$$

This means that $\xi \in B'_\gamma$, where $\gamma \equiv \sigma^{-1}\tau \pmod{\delta^{-1}}$. The lemma is proved. □

10.6 Basic Domains

Lemma 10.2. *We have*

$$\int_{B'_\gamma} S(\xi, T)E(-\chi\xi)\,dx = JG(\gamma)E(-\chi\gamma)T^{2(s-k)r}a^{-2r} + O(T^{2(s-k)r}a^{-2r-7z}),$$

where

$$J = D^{\frac{1}{2}(1-s)}k^{-2sr}|N(\gamma_1\cdots\gamma_s)|^{-1/k}\prod_m Q_m, \qquad \gamma_i = \alpha_i/a, \quad 1 \leq i \leq s,$$

and

$$Q_m = \int_{U_m}\prod_{i=1}^{s} u_i^{\frac{1}{k}-1}\,du\,d\varphi$$

in which $du = \prod_{1\leq i<s} du_i$, $d\varphi = \prod_{1\leq i<s} d\varphi_i$ *and* U_m *denotes the domain:*

$$\begin{aligned}&0 \leq u_i \leq |\gamma_i^{(m)}|^2 \quad (1 \leq i < s), \qquad -\pi \leq \varphi_j \leq \pi \quad (1 \leq j < s-1),\\ &u_s = \left|u_1^{\frac{1}{2}}e^{i\varphi_1} + \cdots + u_{s-1}^{\frac{1}{2}}e^{i\varphi_{s-1}}\right|^2.\end{aligned}$$

For the proof of this lemma we first establish:

Lemma 10.3. *We have*

$$\int_{B'_\gamma} S(\xi, T)E(-\chi\xi)\,dx = G(\gamma)E(-\chi\gamma)\int_{E_{2r}} I(\zeta, T)\,dx + O(T^{2(s-k)r}a^{-2r-7z}).$$

Proof. By Lemmas 5.3 and 5.6, we have

$$I_j(\zeta, T) \ll \prod_{i=1}^{2r} \min\left(T,\, a^{-\frac{1}{k}}|\zeta^{(i)}|^{-\frac{1}{k}}\right), \qquad 1 \le j \le s,$$

and

$$S_j(\xi, T) = G_j(\gamma) I_j(\zeta, T) + O\left(T^{2r-1} a^{\frac{z}{2r}}\right).$$

Therefore

$$S(\xi, T) = G(\gamma) I(\zeta, T) + O\left(T^{2sr-1} a^{\frac{z}{2r}}\right).$$

By (5.4) with $\alpha = 1$ we obtain

$$\int_{B'_\gamma} dx \ll (ht)^{-2r} = a^{-2r-2kry+z/3},$$

and by (10.6), $(\frac{1}{3} + \frac{1}{2r})z - y < -7z$. Therefore

$$\int_{B'_\gamma} S(\xi, T) E(-\chi\xi)\, dx = G(\gamma) E(-\chi\gamma) \int_{B'_\gamma} I(\zeta, T) E(-\chi\zeta)\, dx \tag{10.32}$$
$$+ O\left(T^{2(s-k)r} a^{-2r-7z}\right).$$

If we replace $E(-\chi\zeta)$ by 1 in the integral on the right hand side here, then the error involved is

$$T^{2sr} \int_{B'_\gamma} \|\chi\zeta\|\, dx \ll T^{2sr} H (ht)^{-2r-1} \ll T^{2(s-k)r} a^{-2r-7z}.$$

Therefore (10.32) becomes

$$\int_{B'_\gamma} S(\xi, T) E(-\chi\xi)\, dx = G(\gamma) E(-\chi\gamma) \int_{B'_\gamma} I(\zeta, T)\, dx \tag{10.33}$$
$$+ O\left(T^{2(s-k)r} a^{-2r-7z}\right).$$

If $\mathbf{x}$ is a point in $E_{2r} - B'_\gamma$, then the inequality $h|\zeta^{(i)}| \ge t^{-1}$ is true for at least one index i. Since $s \ge c_{12}$ $(> 30kr)$, it follows from (5.4) and (5.6) that

$$\begin{aligned}
\int_{E_{2r}-B'_\gamma} I(\zeta, T)\, dx &\ll \int_{E_{2r}-B'_\gamma} \prod_{i=1}^{2r} \min\left(T,\, a^{-\frac{1}{k}}|\zeta^{(i)}|^{-\frac{1}{k}}\right)^s dx \\
&\ll \left(\int_{-\pi}^{\pi} \int_{(ht)^{-1}}^{\infty} a^{-\frac{2s}{k}} u^{-\frac{2s}{k}+1} du\, d\varphi\right) \times \\
&\quad \times \left(\int_{-\pi}^{\pi} \int_0^{\infty} \min\left(T^{2s},\, a^{-\frac{2s}{k}} v^{-\frac{2s}{k}}\right) v\, dv\, d\varphi\right)^{r-1} \\
&\ll a^{-\frac{2s}{k}} (ht)^{\frac{2s}{k}-2} a^{-2(r-1)} T^{2(s-k)(r-1)} \\
&\ll T^{2(s-k)r} a^{-\frac{2s}{k} + (\frac{2s}{k}-2)(1+ky-\frac{z}{6r}) - 2r + 2 - 2(s-k)y} \\
&\ll T^{2(s-k)r} a^{-2r-\frac{z}{6r}(\frac{2s}{k}-2)} \ll T^{2(s-k)r} a^{-2r-7z}.
\end{aligned}$$

The lemma follows by substituting this into (10.33). □

Proof of Lemma 10.2. By (5.11) we have

$$\int_{E_{2r}} I(\zeta, T)\, dx = T^{2(s-k)r} J(0).$$

Hence it remains to show that $J(0) = Ja^{-2r}$. But this follows immediately from Theorem 6.2 with the transformation $w_i = a^2 u_i$ $(1 \le i \le s)$ for the integrals H_m. □

10.7 Proof of Theorem 10.1

We have

$$\sum_{\gamma \in \Gamma(t)} 1 \ll \sum_{N(\mathfrak{a}) \le t^{2r}} N(\mathfrak{a}) \ll \sum_{d \le t^{2r}} d^2 \ll t^{6r} = a^{2z}$$

and, by Lemma 3.2,

$$H^{2r} \ll \sum_{\chi \in P(H)} 1 \ll H^{2r}.$$

If (10.21) holds, then, by (10.20) and Lemma 10.2,

$$\begin{aligned} R &= \sum_{\gamma \in \Gamma(t)} \int_{B'_\gamma} F(\xi)\, dx + O\big(H^{2r} T^{2(s-k)r} a^{-2r-4z}\big) \\ &= J\mathfrak{S}(t, H) T^{2(s-k)r} a^{-2r} + O\big(H^{2r} T^{2(s-k)r} a^{-2r-4z}\big), \end{aligned}$$

where

$$\mathfrak{S} = \mathfrak{S}(t, H) = \sum_{\chi \in P(H)} \sum_{\gamma \in \Gamma(t)} G(\gamma) E(-\chi\gamma).$$

Similarly to the proof of Theorem 9.1, we have $\mathfrak{S} \gg H^{2r}$; see §9.8. Since $J > c_{15} > 0$ by (10.16), we have

$$R > c_{16} H^{2r} T^{2(s-k)r} a^{-2r} > 1$$

if $a > c_6(k, \mathbb{K}, x'')$. The theorem is proved. □

Notes

Theorem 10.1 is an analogue of a theorem due to Schmidt; see Wang [2]. He first proved that if $a_1, \ldots, a_s$ are rational integers and $s \ge c_{17}(k, \epsilon)$ then

$$a_1 x_1^k + \cdots + a_s x^k = 0$$

has a solution $\mathbf{x}$ satisfying

$$\max |x_i| \le \max(1, |a_i|)^\epsilon,$$

where $x_i \in \mathbb{Z}$ $(1 \le i \le s)$ if k is odd, where $x_i \in \mathbb{Z}$ or $x_i \in e^{\pi i/k}\mathbb{Z}$ if k is even. If $\mathbb{K}$ has real conjugates, then Theorem 10.1 holds if the conclusion $\boldsymbol{\lambda} \in J^s$ is replaced by $\lambda_j \in J$ or $\lambda_j \in e^{\pi i/k} J$ $(1 \le j \le s)$; see Wang [2].

Chapter 11

Diophantine Inequalities for Forms

11.1 Introduction

A form $F(\boldsymbol{\lambda})$ of degree k can be written as

$$F(\boldsymbol{\lambda}) = \sum_{1 \le i_1, \dots, i_k \le s} a(i_1, \dots, i_k) \lambda_{i_1} \cdots \lambda_{i_k}$$

with $a(i_1, \dots, i_k)$ a symmetric function of its arguments. With $F(\boldsymbol{\lambda})$ we associate the multilinear form

$$\hat{F}(\boldsymbol{\lambda}_1, \dots, \boldsymbol{\lambda}_k) = \sum_{1 \le i_1, \dots, i_k \le s} a(i_1, \dots, i_k) \lambda_{i_1}(1) \cdots \lambda_{i_k}(k).$$

Here $\boldsymbol{\lambda}_i = (\lambda_1(i), \dots, \lambda_s(i))$, $1 \le i \le k$. Note that $\hat{F}$ is linear in each vector $\boldsymbol{\lambda}_i$, symmetric in the k vectors $\boldsymbol{\lambda}_1, \dots, \boldsymbol{\lambda}_k$, and that

$$F(\boldsymbol{\lambda}) = \hat{F}(\boldsymbol{\lambda}, \dots, \boldsymbol{\lambda}).$$

Denote by $|F|$ the maximum absolute value of the coefficients of $F(\boldsymbol{\lambda})$.

The object of this chapter is to prove the following:

Theorem 11.1. *Given positive integers h, q and $k_1, \dots, k_h$, and a positive number E, however large, there is a constant*

$$c_1 = c_1(k_1, \dots, k_h; r, q, E)$$

with the following properties. Let $T \ge 1$ and $F_1(\boldsymbol{\lambda}), \dots, F_h(\boldsymbol{\lambda})$ be forms in $\boldsymbol{\lambda} = (\lambda_1, \dots, \lambda_s)$, where $s \ge c_1$, with complex coefficients and of degrees $k_1, \dots, k_h$ respectively. Then there are q linearly independent points $\boldsymbol{\lambda}(1), \dots, \boldsymbol{\lambda}(q)$ in J^s with

$$\|\boldsymbol{\lambda}(i)\| \le T, \qquad 1 \le i \le q$$

and

$$|\hat{F}_j(\boldsymbol{\lambda}(i_1), \dots, \boldsymbol{\lambda}(i_{k_j}))| \ll T^{-E} |F_j|, \qquad 1 \le j \le h, \quad 1 \le i_1, \dots, i_{k_j} \le q.$$

In the theorem, and in the following, the constants implicit in $\ll$ and O may depend on $k_1, \ldots, k_h, \mathbb{K}, q, E$ and ϵ, but not on $T, F_1, \ldots, F_h$ and G. As a particular case of the theorem we have

$$|F_j(\boldsymbol{\lambda}(i))| \ll T^{-E}|F_j|, \qquad 1 \leq j \leq h, \quad 1 \leq i \leq q.$$

Suppose now that $G_1(\boldsymbol{\lambda}), \ldots, G_h(\boldsymbol{\lambda})$ are forms of degrees $k_1, \ldots, k_h$ respectively, and with coefficients in J. Let $\|G_i\|$ denote the maximum absolute value of its coefficients and the conjugates. Further let

$$G = \max_i \left(1, \|G_i\|\right).$$

Suppose that $s \geq c_1(k_1, \ldots, k_h; r, q, 4rk_1 \cdots k_h/\epsilon) = c_2(k_1, \ldots, k_h; r, q, \epsilon)$ say, where $0 < \epsilon < 1$. We apply Theorem 11.1 to $G_i(\boldsymbol{\lambda})\,(1 \leq i \leq h)$ with $T = T_0 G^\epsilon$, where $T_0 = T_0(k_1, \ldots, k_h; r, q, \epsilon)$ is to be chosen later. We obtain q linearly independent points $\boldsymbol{\lambda}(1), \cdots, \boldsymbol{\lambda}(q)$ in J^s with

$$\|\boldsymbol{\lambda}(i)\| \leq T_0 G^\epsilon, \qquad 1 \leq i \leq q$$

and

$$\text{(11.1)} \quad \left|\hat{G}_j(\boldsymbol{\lambda}(i_1), \ldots, \boldsymbol{\lambda}(i_{k_j}))\right| \ll GT^{-4rk_1 \cdots k_h/\epsilon} \ll T_0^{-4rk_1 \cdots k_h} G^{-4rk_1 \cdots k_h + 1}.$$

On the other hand, $k_j!\, \hat{G}_j(\boldsymbol{\lambda}(i_1), \ldots, \boldsymbol{\lambda}(i_{k_j}))$ is an integer in $\mathbb{K}$ and

$$\|k_j!\, \hat{G}_j(\boldsymbol{\lambda}(i_1), \ldots, \boldsymbol{\lambda}(i_{k_j}))\| \ll GT^{k_j} \ll T_0^{k_j} G^{k_j+1}.$$

If $\hat{G}_j(\boldsymbol{\lambda}(i_1), \ldots, \boldsymbol{\lambda}(i_{k_j})) \neq 0$, then

$$\left|N\left(k_j!\hat{G}_j(\boldsymbol{\lambda}(i_1), \ldots, \boldsymbol{\lambda}(i_{k_j}))\right)\right| \geq 1,$$

and therefore

$$\left|\hat{G}_j(\boldsymbol{\lambda}(i_1), \ldots, \boldsymbol{\lambda}(i_{k_j}))\right| \gg \left(T_0^{k_j} G^{k_j+1}\right)^{-2r+1} \gg T_0^{-2rk_j} G^{-4rk_j+1},$$

which leads to a contradiction with (11.1) if T_0 is large. Therefore

$$\hat{G}_j(\boldsymbol{\lambda}(i_1), \ldots, \boldsymbol{\lambda}(i_{k_j})) = 0,$$

and we have the following:

Theorem 11.2. *Given positive integers $k_1, \ldots k_h$ and q, together with a positive ϵ, however small, there is a constant $c_2 = c_2(k_1, , \ldots, k_h; r, q, \epsilon)$ with the following property: Let $G_1(\boldsymbol{\lambda}), \ldots, G_h(\boldsymbol{\lambda})$ be forms in $\boldsymbol{\lambda} = (\lambda_1, \ldots, \lambda_s)$, where $s \geq c_2$, with respective degrees $k_1, \ldots, k_h$, and with coefficients in J. Then $G_1(\boldsymbol{\lambda}), \ldots, G_h(\boldsymbol{\lambda})$ vanish on a q-dimensional subspace spanned by q points $\boldsymbol{\lambda}(1), \ldots, \boldsymbol{\lambda}(q)$ in J^s having*

$$\|\boldsymbol{\lambda}(i)\| \ll G^\epsilon, \qquad 1 \leq i \leq q.$$

Remark.. The conclusion of Theorem 11.2 is still true with a suitable definition on G, namely that if $G_1, \ldots, G_h$ are forms with coefficients in integers in any given algebraic number field $\mathbb{K}_1$ of degree n_1 provided only that $s \geq c_2'(k_1, \ldots, k_h; rn_1, q, \epsilon)$.

In fact, let $\mathbb{K}^\star$ be the minimum field containing both $\mathbb{K}$ and $\mathbb{K}_1$. Let $w_1^\star, \ldots, w_t^\star$ be an integral basis of $\mathbb{K}^\star$ with respect to $\mathbb{K}$. Then $t \leq n_1$ and G_i can be written as

$$G_i = G_{i1}w_1^\star + \cdots + G_{it}w_t^\star, \qquad 1 \leq i \leq h,$$

where $G_{ij}\,(1 \leq i \leq h,\, 1 \leq j \leq t)$ are forms with coefficients in integers of $\mathbb{K}$. Therefore we may consider the system of forms G_{ij} instead of $G_i\,(1 \leq i \leq h)$, and the assertion follows.

Another apparently more general formulation of Theorem 11.1 is that $F_1, \ldots, F_h$ are any polynomials without constant terms and with respective degrees $k_1, \ldots, k_h$ in $\boldsymbol{\lambda}$.

11.2 A Single Additive Form

We first proof a very special case of Theorem 11.1, namely when there is only one additive form.

Lemma 11.1. *Given $k \geq 1$ and E, however large, there is a constant $c_3(k, r, E)$ with the following property: Let*

$$A(\boldsymbol{\lambda}) = \sum_{i=1}^{s} \alpha_i \lambda_i^k, \qquad s \geq c_3,$$

be an additive form of degree k with complex coefficients. Then, for real $T \geq 1$, there is a nonzero point $\boldsymbol{\lambda} \in J^s$ with

$$\|\boldsymbol{\lambda}\| \leq T \qquad \text{and} \qquad |A(\boldsymbol{\lambda})| \ll T^{-E}|A|,$$

where the constant implicit in $\ll$ depends on $k, \mathbb{K}$ and E.

If α_i are integers, then Theorem 10.1 follows from Lemma 11.1 at once; see §11.1. On the other hand, the proof of Lemma 11.1 depends on Theorem 10.1.

Suppose that we can prove the conclusion of Lemma 11.1 for $s \geq c_3$ and $T \geq c_4(k, \mathbb{K}, E)$. Then, for $1 \leq T \leq c_4$, we put

$$\lambda_1 = 1, \qquad \lambda_i = 0, \quad 2 \leq i \leq s,$$

so that

$$\|\boldsymbol{\lambda}\| = 1 \qquad \text{and} \qquad |A(\boldsymbol{\lambda})| \leq |A| = c_4^E c_4^{-E}|A| \ll T^{-E}|A|.$$

Therefore we need only establish the lemma for large values of T, say $T \geq c_4$.

Choose a positive ϵ so small that

$$2\epsilon + \frac{E+2\epsilon}{E+1/8} < 1.$$

The lemma is obvious if there is an α_i with $|\alpha_i| \leq T^{-E}|A|$. In fact, we may take $\lambda_i = 1$, $\lambda_j = 0\,(j \neq i)$ in this case. So we may suppose that

$$T^{-E}|A| < |\alpha_i| \leq |A|, \qquad 1 \leq i \leq s.$$

Cover the interval $[-E, 0]$ by a finite number of intervals I with length ϵ. For one of these I, at least $\left[s/([E/\epsilon]+1)\right]$ of the α_i are of the type $|\alpha_i| = T^{a_i}|A|$ with $a_i \in I$. We may suppose without loss of generality that this holds for all i. Put

$$L = T^{\epsilon}|A| = T^{\epsilon} \max |\alpha_i|$$

and choose rational integers $q_1, \ldots, q_s$, each as large as possible, with

$$|\alpha_i| q_i^k \leq L, \qquad 1 \leq i \leq s.$$

Since $L/|\alpha_i| \geq T^{\epsilon}$, we have

$$\tfrac{1}{2}L \leq |\alpha_i| q_i^k \leq L \tag{11.2}$$

if T is large, and also

$$q_i^k \leq \frac{L}{|\alpha_i|} = \frac{T^{\epsilon}|A|}{|\alpha_i|} \leq T^{2\epsilon}.$$

Suppose that Lemma 11.1 is true for

$$B(\boldsymbol{\mu}) = \alpha_1 q_1^k \mu_1^k + \cdots + \alpha_s q_s^k \mu_s^k$$

with

$$T_0 = T^{\frac{E+2\epsilon}{E+1/8}}, \qquad E_0 = E + \tfrac{1}{8}$$

in place of T and E; that is, there is a nonzero point $\boldsymbol{\mu} \in J^s$ with

$$\|\boldsymbol{\mu}\| \leq T_0 \qquad \text{and} \qquad |B(\boldsymbol{\mu})| \ll T_0^{-E_0}|B|.$$

Let $\lambda_i = q_i \mu_i$, $1 \leq i \leq s$. Since $|B| \leq L = T^{\epsilon}|A|$, we have

$$\|\boldsymbol{\lambda}\| \ll (\max q_i)\|\boldsymbol{\mu}\| \ll T^{2\epsilon + \frac{E+2\epsilon}{E+1/8}} < T$$

and

$$|A(\boldsymbol{\lambda})| \ll T^{-(\frac{E+2\epsilon}{E+1/8})(E+\frac{1}{8})} T^{\epsilon}|A| \ll T^{-E}|A|,$$

if T is large; in other words Lemma 11.1 is true for $A(\boldsymbol{\lambda})$.

The special point about $B(\boldsymbol{\mu})$ is that, by (11.2), each of its coefficients is at least $\frac{1}{2}|B|$ in absolute value. It is clear that if Lemma 11.1 is true with $E+\frac{1}{8}$ in place of E for the form $A(\boldsymbol{\lambda})$ with

$$\tfrac{1}{2}|A| \le |\alpha_i| \le |A|, \qquad 1 \le i \le s,$$

then it is true with E for general forms. By homogeneity we may replace the above relation by

$$|A| = 1 \quad \text{and} \quad \tfrac{1}{2} \le |\alpha_i| \le 1, \qquad 1 \le i \le s. \tag{11.3}$$

It will suffice to prove the following statements.

1) The conclusion of Lemma 11.1 is true for $0 \le E \le \frac{1}{2}$, for forms $A(\boldsymbol{\lambda})$ with (11.3), provided only that $s \ge c_5(k, r, E)$.

2) The conclusion of Lemma 11.1 is true for E, for forms $A(\boldsymbol{\lambda})$ with (11.3), provided only that $s \ge c_5(k, r, E)$ and that Lemma 11.1 is true for $E - \frac{1}{4}$ for general additive forms.

11.3 A Variant Circle Method

Given a set of complex numbers $C_1, \ldots, C_{2r}$ with $C_{m+r} = \bar{C_m}$, there is a unique set of real numbers $y_1, \ldots, y_{2r}$ such that

$$C_m = y_1 w_1^{(m)} + \cdots + y_{2r} w_{2r}^{(m)}.$$

Suppose that α_i are complex number satisfying (11.3). We define, for each i,

$$\alpha_i^{(m)} = \alpha_i, \qquad \alpha_i^{(m+r)} = \bar{\alpha_i}.$$

Then α_i has the unique representation

$$\alpha_i^{(m)} = z_1 w_1^{(m)} + \cdots + z_{2r} w_{2r}^{(m)}, \qquad z_i \in \mathbb{R}, \qquad 1 \le i \le 2r.$$

Set

$$K(x) = \chi_{2^{-1}T^{-E}}(x)^2 = \left(\frac{\sin \pi x\, T^{-E}}{\pi x}\right)^2 \quad \text{and} \quad K(\mathbf{x}) = \prod_{i=1}^{2r} K(x_i),$$

where x is a real variable and $\mathbf{x}$ is a real vector variable.

We wish to estimate the number W of solutions of the inequality

$$|A(\boldsymbol{\lambda})| \ll T^{-E} \tag{11.4}$$

in $\boldsymbol{\lambda} \in J^s$ subject to

$$\|\boldsymbol{\lambda}\| \le T. \tag{11.5}$$

Our plan is to show that either $W > 1$ or else 1) and 2) hold, and therefore 1) and 2) hold in every case. Put

$$\theta = \bigl(rk + (r+1)E + 3\bigr)^{-1}(4k + 5E)^{-1}, \tag{11.6}$$

$$u = \begin{cases} 1, & \text{in the case 1)}, \\ c_3(k,r,E-\frac{1}{4}), & \text{in the case 2)}, \end{cases} \tag{11.7}$$

$$v = c_6(k,r,\theta), \tag{11.8}$$

$$q = uv, \tag{11.9}$$

and choose $\epsilon_1 > 0$ so small that

$$41Gqr\epsilon_1 < 1. \tag{11.10}$$

Here $c_6(k,r,\theta)$ is a constant such that, if $s \geq c_6$, then, given any additive form $A(\boldsymbol{\lambda})$ with integral coefficients, there is a nonzero point $\boldsymbol{\lambda} \in J^s$ with

$$A(\boldsymbol{\lambda}) = 0 \qquad \text{and} \qquad \|\boldsymbol{\lambda}\| \ll A^\theta;$$

see Theorem 10.1. Now let s be so large that

$$\big(2r(r+1)(k+E)+2r+2\big)(s-q+1)^{-1} < \epsilon_1. \tag{11.11}$$

In what follows, the constants in $\ll$ or O may depend on s, in addition to $k, \mathbb{K}$ and E. However, observe that if Lemma 11.1 holds for a particular value of s, then it also holds for larger values of s. We assume T to be large, say $T \geq c_4$.

Lemma 11.2. *We have, for real* Q,

$$T^E \int_{-\infty}^{\infty} e(xQ)K(x)\,dx = \begin{cases} 1 - T^E|Q|, & \textit{if } |Q| < T^{-E}, \\ 0, & \textit{if } |Q| \geq T^{-E}. \end{cases} \tag{11.12}$$

Proof. Since

$$\begin{aligned} \int_{-T}^{T} \left(\frac{\sin \pi x}{\pi x}\right)^2 dx &= -\frac{\sin^2 \pi x}{\pi^2 x}\Bigg|_{-T}^{T} + \frac{1}{\pi^2}\int_{-T}^{T} \frac{2\pi \sin \pi x \cos \pi x}{x}\,dx \\ &= \frac{1}{\pi}\int_{-T}^{T} \frac{\sin 2\pi x}{x}\,dx + O\left(\frac{1}{T}\right), \end{aligned}$$

we have

$$\int_{-\infty}^{\infty} \left(\frac{\sin \pi x}{\pi x}\right)^2 dx = 1, \quad \text{and hence} \quad \int_{-\infty}^{\infty} \left(\frac{\sin \pi Q x}{\pi x}\right)^2 dx = |Q|.$$

This gives

$$\begin{aligned} \int_{-\infty}^{\infty} e(xQ)\left(\frac{\sin \pi x}{\pi x}\right)^2 dx &= \int_{-\infty}^{\infty} \cos 2\pi x Q \left(\frac{\sin \pi x}{\pi x}\right)^2 dx \\ &= \frac{1}{2}\int_{-\infty}^{\infty} \frac{\sin^2 \pi x(Q+1) + \sin^2 \pi x(Q-1) - 2\sin^2 \pi x Q}{\pi^2 x^2}\,dx \\ &= \frac{1}{2}\big(|Q+1| + |Q-1| - 2|Q|\big). \end{aligned}$$

Making the substitutions $x' = T^E x$ and $T^E Q$ for Q, and write x' as x again, we see that the lemma is proved. □

Define $S_i(\xi, T)$, $I_i(\xi, T)\,(1 \le i \le s)$, $S(\xi, T)$ and $I(\xi, T)$ as in §5.1. Expand $\sum_{i=1}^{s} \alpha_i \lambda_i^k$ as

$$\sum_{i=1}^{s} \alpha_i^{(m)} \lambda_i^{(m)k} = A_1 w_1^{(m)} + \cdots + A_{2r} w_{2r}^{(m)}. \tag{11.13}$$

Then, by Lemma 11.2, we have

$$\begin{aligned} T^{2rE} \int_{E_{2r}} & S(\xi, T) K(\mathbf{x})\, dx \\ &= T^{2rE} \sum_{\lambda_1 \in P(T)} \cdots \sum_{\lambda_s \in P(T)} \prod_{i=1}^{2r} \int_{-\infty}^{\infty} e(x_i A_i) K(x_i)\, dx_i \\ &= \sum_{\substack{\lambda_1 \in P(T) \\ |A_i| < T^{-E}}} \cdots \sum_{\lambda_s \in P(T)} (1 - |A_1| T^E) \cdots (1 - |A_{2r}| T^E). \end{aligned} \tag{11.14}$$

Let

$$\eta_i = y_{i1} w_1 + \cdots + y_{i,2r} w_{2r}, \qquad dY_i = dy_{i1} \cdots dy_{i,2r}$$

and

$$\sum_{i=1}^{s} \alpha_i^{(m)} \eta_i^{(m)k} = B_1 w_1^{(m)} + \cdots B_{2r} w_{2r}^{(m)}.$$

Then, by Lemma 11.2, we have

$$\begin{aligned} T^{2rE} \int_{E_{2r}} & I(\xi, T) K(\mathbf{x})\, dx \\ &= \int_{\substack{P(T) \\ |B_i| < T^{-E}}} \cdots \int_{P(T)} (1 - |B_1| T^E) \cdots (1 - |B_{2r}| T^E)\, dY_1 \cdots dY_s. \end{aligned} \tag{11.15}$$

If $|A_i| < T^{-E}\,(1 \le i \le 2r)$, then it follows from (11.13) that $A(\boldsymbol{\lambda}) \ll T^{-E}$, and thus the right hand side of (11.14) gives a lower bound for W. The general ider will be to show that the right hand side of (11.15) is large and that the left hand sides of (11.14) and (11.15) differ little.

This method of treating diophantine inequalities was first appeared in the paper of Davenport and Heilbronn [1]. We call it the variant circle method of Davenport and Heilbronn.

11.4 Continuation

Lemma 11.3. *The righ hand side of* (11.15) *is* $\gg T^{2(s-k-E)r}$.

Proof. Let

$$\eta_j^{(m)} = u_{jm}^{\frac{1}{k}} e^{i\varphi_{jm}/k}, \qquad 1 \le j \le s.$$

The Jacobian of y_{ji} $(1 \le i \le 2r)$ with respect to u_{jm}, φ_{jm} is

$$2^r k^{-2r} D^{-\frac{1}{2}} \prod_m u_{jm}^{\frac{2}{k}-1},$$

and so the right hand side of (11.15) is equal to

$$2^{sr} k^{-2sr} D^{-\frac{s}{2}} \Phi, \tag{11.16}$$

where

$$\Phi = \int_U \prod_{i=1}^{2r} (1 - T^E |B_i|) \prod_{j=1}^{s} \prod_m u_{jm}^{\frac{2}{k}-1} du\, d\varphi, \tag{11.17}$$

in which $du = \prod_{1 \le j \le s} \prod_m du_{jm}$, $d\varphi = \prod_{1 \le j \le s} \prod_m d\varphi_{jm}$ and U denotes the domain:

$$0 \le u_{jm} \le T^k, \quad -\pi \le \varphi_{jm} \le \pi \ (1 \le j \le s), \qquad |B_i| < T^{-E} \ (1 \le i \le 2r).$$

Let

$$u_{jm} = |\alpha_j|^{-1} T^{-E} v_{jm}, \qquad 1 \le j \le s.$$

Then

$$\Phi = |\alpha_1 \cdots \alpha_s|^{-2r/k} T^{-2Esr/k} \Psi, \tag{11.18}$$

where

$$\Psi = \int_V \prod_{i=1}^{2r} (1 - T^E |B_i|) \prod_{j=1}^{s} \prod_m v_{jm}^{\frac{2}{k}-1} dv\, d\varphi, \tag{11.19}$$

in which $dv = \prod_{1 \le j \le s} \prod_m dv_{jm}$ and V denotes the domain:

$$0 \le v_{jm} \le |\alpha_j| T^{E+k}, \ -\pi \le \varphi_{jm} \le \pi \ (1 \le j \le s), \ \ |B_i| < T^{-E} \ (1 \le i \le 2r).$$

Let

$$\theta_{jm} = \varphi_{jm} + \arg \alpha_j^{(m)}, \ 1 \le i \le s \qquad \text{and} \qquad T^E B_i = D_i, \ 1 \le i \le 2r.$$

Then

$$\sum_{j=1}^{s} v_{jm}e^{i\theta_{jm}} = T^E \sum_{j=1}^{s} \alpha_j^{(m)} u_{jm} e^{i\varphi_{jm}} = T^E \sum_{j=1}^{s} \alpha_j^{(m)} \eta_j^{(m)k}$$

$$= D_1 w_1^{(m)} + \cdots + D_{2r} w_{2r}^{(m)}$$

and

$$\Psi = \int_W \prod_{i=1}^{2r} (1 - |D_i|) \prod_{j=1}^{s} \prod_m v_{jm}^{\frac{2}{k}-1} dv\, d\theta,$$

where $d\theta = \prod_{1 \le j \le s} \prod_m d\theta_{jm}$ and W denotes the domain:

$$0 \le v_{jm} \le |\alpha_j| T^{E+k}, \ -\pi \le \theta_{jm} \le \pi \ (1 \le j \le s), \quad |D_i| < 1 \ (1 \le i \le 2r).$$

Since $\frac{1}{2} \le |\alpha_i| \le 1$, $1 \le i \le s$, the domain W':

$$\tfrac{1}{8}T^{E+k} \le v_{1m} \le \tfrac{1}{4}T^{E+k}, \quad \frac{T^{E+k}}{32s} \le v_{jm} \le \frac{T^{E+k}}{16s}, \quad -\pi \le \theta_{jm} \le \pi,$$

$$2 \le j \le s-1, \quad |D_i| < \frac{1}{2}, \ 1 \le i \le 2r, \quad \sum_{j=1}^{s} v_{jm} e^{i\theta_{jm}} = \sum_{i=1}^{2r} D_i w_i^{(m)}$$

is contained in W. The volume of W' is $\gg T^{(E+k)(s-2)r}$. In fact, for any given v_{jm}, θ_{jm} $(1 \le j \le s-1)$ in W', we let

$$v_{1m}e^{i\theta_{1m}} + \cdots + V_{s-1,m}e^{i\theta_{s-1,m}} = ve^{i\theta}.$$

Then $T^{E+k}/16 \le v \le 5T^{E+k}/16$, and v_{sm}, θ_{sm} satisfy $1 \ll v_{sm} - v \ll 1$ and $\theta_{sm} - \theta \ll T^{-E-k}$. The assertion follows. The integrand for the integral Ψ in W' is $\gg T^{(E+k)(\frac{2}{k}-1)sr}$. Hence

$$\Psi \gg T^{(E+k)(s-2)r+(E+k)(\frac{2}{k}-1)sr} \gg T^{2(s-k-E)r+2Esr/k},$$

and the lemma follows from (11.16) – (11.19). □

Lemma 11.4. *We have*

$$\int_{\|\xi\| \le T^{\frac{9}{10}-k}} S(\xi,T)K(\mathbf{x})\, dx \gg T^{2(s-k-2E)r}. \tag{11.20}$$

Proof. Take $t = 2$ and $h = 2^{-1}T^{k-9/10}$. Then B_0' is the set

$$\{\mathbf{x} : \mathbf{x} \in U_n, \ \|\xi - \gamma_0\| \le T^{\frac{9}{10}-k}, \text{ for some } \gamma_0 \equiv 0 \ (\mathrm{mod}\ \delta^{-1})\};$$

see §9.5. By Lemma 5.6,

$$\sum_{\lambda P(T)} E(\alpha_i \xi \lambda^k) = \int_{P(T)} E(\alpha_i \xi \eta^k)\, dy + O(T^{2r-1/10}). \tag{11.21}$$

By (5.4), (5.6) and Lemma 5.3, we have, for $s > 2k$,

$$\begin{aligned}
&\int_{|\xi^{(m)}|>T^{\frac{9}{10}-k}} I(\xi,T)K(\mathbf{x})\,dx \\
&\ll T^{-4rE}\int_{|\xi^{(m)}|>T^{\frac{9}{10}-k}} \prod_{i=1}^{2r}\min\left(T^s,|\xi^{(i)}|^{-s/k}\right)dx \\
&\ll T^{-4rE}\int_{T^{\frac{9}{10}-k}}^{\infty}\int_{-\pi}^{\pi} u^{-\frac{2s}{k}+1}du\,d\varphi\left(\int_0^{\infty}\int_{-\pi}^{\pi}\min\left(T^{2s},w^{-s/k}\right)w\,dw\,d\theta\right)^{r-1} \\
&\ll T^{-4rE+(\frac{9}{10}-k)(-\frac{2s}{k}+2)+2(s-k)(r-1)} \\
&\ll T^{2(s-k-2E)r-\frac{9s}{5k}+\frac{9}{5}} \le T^{2(s-k-2E)r-1}.
\end{aligned}$$

Therefore it follows from Lemma 11.3 that

$$\int_{\|\xi\|\le T^{\frac{9}{10}-k}} I(\xi,T)K(\mathbf{x})\,dx \gg T^{2(s-k-2E)r}. \tag{11.22}$$

It remains to compare the integral (11.22) with the one in (11.20). By Lemma 5.3 and (11.21) we have, for $\|\xi\|\le T^{\frac{9}{10}-k}$,

$$\begin{aligned}
|S(\xi,T)-T(\xi,T)| &= \Big|\prod_{i=1}^{s}\left(I_i(\xi,T)+O(T^{2r-1/10})\right)-I(\xi,T)\Big| \\
&\ll T^{2r-1/10}\prod_{i=1}^{2r}\min\left(T^{s-1},\,|\xi^{(i)}|^{-\frac{s-1}{k}}\right)+T^{s(2r-1/10)},
\end{aligned}$$

and therefore the left hand sides of (11.20) and (11.22) differ by

$$\begin{aligned}
&\ll T^{-4rE+2r-1/10}\left(\int_0^{T^{9/10-k}}\int_{-\pi}^{\pi}\min\left(T^{2(s-1)},\,u^{-\frac{2(s-1)}{k}}\right)u\,du\,d\varphi\right)^r \\
&\qquad\qquad + T^{-4rE}\int_{\|\xi\|\le T^{\frac{9}{10}-k}} T^{s(2r-1/10)}dx \\
&\ll T^{-4rE+2r-1/10}\left(\int_0^{T^{-k}} T^{2s-2}u\,du+\int_{T^{-k}}^{\infty}u^{-\frac{2(s-1)}{k}+1}du\right)^r \\
&\qquad\qquad + T^{2(s-k-2E)r-\frac{s}{10}+\frac{9r}{5}} \\
&\ll T^{2(s-k-2E)r-\frac{1}{10}},
\end{aligned}$$

if $s\ge 20r$. The lemma is proved. □

Lemma 11.5. *We have, for any m,*

$$\int_{|\xi^{(m)}|>T^{kr+(r+1)E+1}} S(\xi,T)K(\mathbf{x})\,dx \ll T^{2(s-k-2E)r-1}.$$

Proof. From $|\xi^{(m)}|>T^{kr+(r+1)E+1}$ we see that at least one of the x_i in the expression $\xi=x_1\rho_1+\dots+x_{2r}\rho_{2r}$ satisfies $|x_i|\gg T^{kr+(r+1)E+1}$. Since

$$|K(x)| \leq \min\left(T^{-2E},\, \pi^{-2}|x|^{-2}\right),$$

the integral in the lemma is

$$\ll T^{2sr}\left(\int_{T^{kr+(r+1)E+1}}^{\infty} \frac{dx}{x^2}\right)^2 \left(\int_0^{\infty} \min\left(T^{-2}E,\, x^{-2}\right) dx\right)^{2r-2}$$
$$\ll T^{2sr-2kr-2(r+1)E-2-2(r-1)E} \ll T^{2(s-k-2E)r-1}.$$

The lemma is proved. □

Let Q_m denote the domain:

$$\|\xi\| \leq T^{kr+(r+1)E+1}, \qquad |\xi^{(m)}| > T^{\frac{9}{10}-k}.$$

Suppose for a moment that the r integrals satisfy

$$\int_{Q_m} |S(\xi, T)|\, dx \leq T^{2(s-k)r-1}. \tag{11.23}$$

Then it follows from Lemmas 11.4 and 11.5 that

$$\int_{E_{2r}} S(\xi, T)K(\mathbf{x})\, dx = \int_{\|\xi\| \leq T^{\frac{9}{10}-k}} S(\xi, T)K(\mathbf{x})\, dx + \left(T^{2(s-k-2E)r-1}\right)$$
$$\gg T^{2(s-k-2E)r},$$

and, in view of (11.14), it follows that

$$W \gg T^{2(s-k-E)r} > 1.$$

11.5 Proof of Lemma 11.1

We may thus suppose that (11.23) is false, so that there is a Q_m such that

$$\int_{Q_m} |S(\xi, T)|\, dx > T^{2(s-k)r-1}. \tag{11.24}$$

Let $\sigma(\mathbf{j})$ be a transformation

$$\gamma^{(m)} \to \gamma^{(j_m)}, \qquad 1 \leq m \leq r,$$

where $\mathbf{j} = (j_1, \ldots, j_r)$ is a permutation of $(1, \ldots, r)$. Since $S(\xi, T)$ is invariant under the group $\{\sigma(\mathbf{j})\}$, the inequality (11.24) is true for any m. In particular it holds for $m = 1$, that is

$$\int_{Q_1} |S(\xi, T)|\, dx > T^{2(s-k)r-1}. \tag{11.25}$$

There is a $\xi = \xi^{(1)}$ satisfying

$$(11.26) \qquad \|\xi\| \le T^{kr+(r+1)E+1}, \qquad |\xi| > T^{\frac{9}{10}-k}$$

and

$$(11.27) \qquad |S(\xi, T)| \ge T^{2(s-(r+1)(E+k)-1)r-2};$$

otherwise we would have

$$\int_{Q_1} |S(\xi, T)|\, dx < T^{2(s-(r+1)(E+k)-1)r-2} \int_{\|\xi\| \le T^{kr+(r+1)E+1}} dx \ll T^{2(s-k)r-2}$$

which contradicts with (11.25) if T is large.

We may suppose without loss of generality that

$$|S_1(\xi, T)| \ge \cdots \ge |S_s(\xi, T)|.$$

The left hand side of (11.27) is

$$\ll |S_q(\xi, T)|^{s-q+1} T^{2(q-1)r},$$

so that, by (11.11),

$$(11.28) \qquad |S_i(\xi, T)| \gg T^{2r - \frac{2r(r+1)(E+k)+2r+2}{s-q+1}} \ge T^{2r-\epsilon_1}, \qquad 1 \le i \le q.$$

In view of (11.10) and (11.11) we can apply Theorem 3.3 with $C = T^{2r-\epsilon_1}$. We obtain integers σ_i, $\beta_i\,(1 \le i \le q)$ with

$$0 < \|\sigma_i\| \ll T^{2G\epsilon_1} \qquad \text{and} \qquad \|\alpha_i \sigma_i \xi - \beta_i\| \ll T^{-k+2G\epsilon_1}, \qquad 1 \le i \le q.$$

Setting $\sigma = \sigma_1 \cdots \sigma_q$ and $\sigma\beta_i/\sigma_i = \tau_i\,(1 \le i \le q)$, we have

$$(11.29) \quad 0 < \|\sigma_i\| \ll T^{2Gq\epsilon_1} \quad \text{and} \quad \|\alpha_i \sigma \xi - \tau_i\| \ll T^{-k+2Gq\epsilon_1}, \quad 1 \le i \le q.$$

Write

$$\alpha_i = \frac{\tau}{\sigma\xi} + \gamma_i, \qquad 1 \le i \le q.$$

Then, by (5.9),

$$|\sigma| \gg T^{-2(2r-1)Gq\epsilon_1},$$

and, by (11.10), (11.26) and (11.29), we have

$$(11.30) \qquad |\gamma_i| \ll |\sigma\xi|^{-1} T^{-k+2Gq\epsilon_1} \ll T^{k-\frac{9}{10}-k+4rGq\epsilon_1} < T^{-4/5}$$

if T is large. So with $\boldsymbol{\lambda} = (\lambda_1, \ldots, \lambda_s) = (\mu_1, \ldots, \mu_q, 0 \ldots, 0) = (\boldsymbol{\mu}, \mathbf{0})$ we have

$$A(\boldsymbol{\lambda}) = (\sigma\xi)^{-1} B(\boldsymbol{\mu}) + C(\boldsymbol{\mu})$$

where

$$B(\boldsymbol{\mu}) = \tau_1 \mu_1^k + \cdots + \tau_q \mu_q^k$$

is a form with coefficients in J and

$$C(\boldsymbol{\mu}) = \gamma_1 \mu_1^k + \cdots + \gamma_q \mu_q^k.$$

From (11.26) and (11.29) we obtain

$$\|\tau_i\| \ll T^{kr+(r+1)E+1+2Gq\epsilon_1}, \qquad 1 \le i \le q,$$

and therefore

$$B < T^{kr+(r+1)E+2} \tag{11.31}$$

by (11.10), if T is large.

1) First, suppose that $0 \le E \le \frac{1}{2}$. Since $q = uv = v = c_6(k, r, \theta)$ by (11.8), (11.8) and (11.9), we find from Theorem 10.1 that there is a nonzero point $\boldsymbol{\mu} \in J^q$ with

$$B(\boldsymbol{\mu}) = 0 \quad \text{and} \quad \|\boldsymbol{\mu}\| \ll B^\theta \ll T^{(kr+(r+1)E+2)\theta} \ll T^{\frac{1}{4k}}$$

by (11.6) and (11.31). With $\boldsymbol{\lambda} = (\boldsymbol{\mu}, \mathbf{0})$ we have, for large T,

$$\|\boldsymbol{\lambda}\| \le T \quad \text{and} \quad |A(\boldsymbol{\lambda})| = |C(\boldsymbol{\mu})| \le qC\|\boldsymbol{\mu}\|^k \ll T^{-4/5}T^{1/4} \le T^{-E}$$

by (11.30).

2) Next, suppose that Lemma 11.1 holds for $E - \frac{1}{4}$ in place of E. We have

$$q = uv, \qquad u = c_6(k, r, E - \tfrac{1}{4}) \quad \text{and} \quad v = c_6(k, r, \theta)$$

by (11.9). Write

$$\begin{aligned} \boldsymbol{\mu} &= (\boldsymbol{\mu}_1, \ldots, \boldsymbol{\mu}_u), \\ B(\boldsymbol{\mu}) &= B_1(\boldsymbol{\mu}_1) + \cdots + B_u(\boldsymbol{\mu}_u), \\ C(\boldsymbol{\mu}) &= C_1(\boldsymbol{\mu}_1) + \cdots + C_u(\boldsymbol{\mu}_u), \end{aligned}$$

where each $\boldsymbol{\mu}_i$ has v components. Since $v = c_6(k, r, \theta)$ by (11.8), there are nonzero points $\boldsymbol{\mu}_1, \ldots, \boldsymbol{\mu}_u$ in J^v with

$$\begin{aligned} &B_i(\boldsymbol{\mu}_i) = 0, \qquad \text{and} \\ &\|\boldsymbol{\mu}_i\| \ll B^\theta \ll T^{(kr+(r+1)E+2)\theta} \ll \min\left(T^{\frac{1}{4k}}, T^{\frac{1}{5E}}\right), \quad 1 \le i \le u. \end{aligned} \tag{11.32}$$

Setting $\boldsymbol{\lambda} = (\nu_1\boldsymbol{\mu}_1, \ldots, \nu_u\boldsymbol{\mu}_u, \mathbf{0})$, we have

$$A(\boldsymbol{\lambda}) = C_1(\boldsymbol{\mu}_1)\nu_1^k + \cdots C_u(\boldsymbol{\mu}_u)\nu_u^k = D(\boldsymbol{\nu}),$$

say. Since $u = c_3(k, r, E - \frac{1}{4})$ and

$$|D| = \max |C_i(\boldsymbol{\mu}_i)| \ll T^{-4/5}T^{1/4} \ll T^{-\frac{1}{2}}$$

by (11.7), (11.30) and (11.32), we have a point $\nu \in J^u$ such that $\|\boldsymbol{\nu}\| \le T^{1-1/4E}$ and

$$|D(\boldsymbol{\nu})| \ll T^{-(1-\frac{1}{4E})(E-\frac{1}{4})}|D| \ll T^{-E+\frac{1}{2}-\frac{1}{16E}-\frac{1}{2}} < T^{-E}$$

if T is large. Consequently, in view of the definition of $\boldsymbol{\lambda}$, we have

$$|A(\boldsymbol{\lambda})| = |D(\boldsymbol{\nu})| < T^{-E},$$

and, by (11.32),

$$\|\boldsymbol{\lambda}\| \ll \|\boldsymbol{\nu}\| \, \|\boldsymbol{\mu}\| \ll T^{1-\frac{1}{4E}+\frac{1}{5E}} < T$$

if T is large. In other words, Lemma 11.1 holds for E for forms with (11.3), and so the lemma is proved. □

11.6 Linear Forms

We prove Theorem 11.1 by induction on $k = \max(k_1, \dots, k_h)$. The case $k = 1$ is established by the box principle. We may suppose that $|F_1| = \cdots = |F_h| = 1$. Let $T \geq 2$, and consider the points $\boldsymbol{\lambda} \in J^s$ with $\lambda_i \in P(T/2)$, $1 \leq i \leq s$; by Lemma 3.2, there are $c_7(\mathbb{K})T^{2sr}$ such points. Given any such $\boldsymbol{\lambda}$, the point

$$\mathbf{p}(\boldsymbol{\lambda}) = \big(F_1(\boldsymbol{\lambda}), \dots, F_h(\boldsymbol{\lambda})\big)$$

lies in the domain $D : (p_1, \dots, p_h)$, where $p_j = x_j + iy_j$, $x_j, y_j \in \mathbb{R}$, $|x_j|, |y_j| \leq sT/2$ for $1 \leq j \leq h$. Let v be the rational integer in

$$\big(c_7 T^{2sr}\big)^{\frac{1}{2h}} - 1 \leq v < \big(c_7 T^{2sr}\big)^{\frac{1}{2h}}$$

and divide the region of p_j in D into v^2 subsquares of side sT/v. Since $v^{2h} < c_7 T^{2sr}$, there are two points $\mathbf{p}(\boldsymbol{\lambda})$ and $\mathbf{p}(\boldsymbol{\lambda}')$ such that $F_j(\boldsymbol{\lambda})$ and $F_j(\boldsymbol{\lambda}')$ both lie in the same subsquare for $j = 1, \dots, h$. Then $\boldsymbol{\Lambda} = \boldsymbol{\lambda} - \boldsymbol{\lambda}'$ satisfies $0 < \|\boldsymbol{\Lambda}\| \leq \|\boldsymbol{\lambda}\| + \|\boldsymbol{\lambda}'\| \leq T$, and, for $j = 1, \dots, h$,

$$|F_j(\boldsymbol{\Lambda})| = |F_j(\boldsymbol{\lambda}) - F_j(\boldsymbol{\lambda}')| \leq \sqrt{\left(\frac{sT}{v}\right)^2 + \left(\frac{sT}{v}\right)^2} < \frac{2sT}{v} \ll T^{-E}$$

if

$$s \geq c_8(h, r, E) = c_1(\underbrace{1, \dots, 1}_{\leftarrow h \rightarrow}; r, 1, E) = u$$

say. So Theorem 11.1 is true for $k = q = 1$. Let $c_1(1, \dots, 1; r, q, E) = qu$. Replace the indices $1 \leq i \leq qu$ by double indices $1 \leq i \leq q$, $1 \leq j \leq u$. If $s \geq qu$, we let

$$\boldsymbol{\lambda}(i) = (\underbrace{0, \dots, 0}_{\leftarrow u(i-1) \rightarrow}, \lambda_{i1}, \dots, \lambda_{iu}, \underbrace{0, \dots, 0}_{\leftarrow s-ui \rightarrow}), \qquad 1 \leq i \leq q.$$

They are linearly independent if $\boldsymbol{\lambda}(i) \neq \mathbf{0}$ for $1 \leq i \leq q$, and, by the case $k = q = 1$, there are $\boldsymbol{\lambda}(i)$ such that

$$\|\boldsymbol{\lambda}(i)\| \leq T \qquad \text{and} \qquad |F_j(\boldsymbol{\lambda}(i))| \ll T^{-E}, \quad 1 \leq j \leq h, \ 1 \leq i \leq q.$$

Therefore Theorem 11.1 holds for $k = 1$.

11.7 A Single Form

Let $F = F(\boldsymbol{\lambda}) = F(\lambda_1, \ldots, \lambda_s)$ and $G = G(\boldsymbol{\mu}) = G(\mu_1, \ldots, \mu_t)$ be two forms. We write

$$F \to G$$

if there are t linearly independent points $\boldsymbol{\lambda}_1, \ldots, \boldsymbol{\lambda}_t$ in J^s with

$$G(\boldsymbol{\mu}) = F(\mu_1 \boldsymbol{\lambda}_1 + \cdots + \mu_t \boldsymbol{\lambda}_t). \tag{11.33}$$

Put

$$\Psi(F, G) = \min \max(\|\boldsymbol{\lambda}_1\|, \ldots, \|\boldsymbol{\lambda}_t\|),$$

where the minimum is over $\boldsymbol{\lambda}_1, \ldots, \boldsymbol{\lambda}_t$ with (11.33). We also set $\Psi(F, G) = \infty$ if $F \not\to G$. The number of variables of F will be denoted by $s(F)$.

Suppose that $k > 1$, and that Theorem 11.1 has already been proved for forms of degree less than k. We use this as induction hypothesis to prove the theorem for a single form of degree k.

Lemma 11.6. *Let $k \geq 1$, $g \geq 1$, $E > 0$ and $T \geq 1$. If F is a form of degree k with $s(F) \geq c_9(k, r, g, E)$, then there is a form G with $F \to G$ and $\Psi(F, G) \leq T$, where G is a form in $g + 1$ variables of the type*

$$G = \sigma \mu^k + F_1(\nu_1, \ldots, \nu_g) + H(\mu, \nu_1, \ldots, \nu_g) \tag{11.34}$$

with

$$|H| \ll T^{-E}|F|.$$

Proof. Let $\mathbf{e}(1), \ldots, \mathbf{e}(g+1)$ be the first $g+1$ unit vectors in E_s. Consider the forms

$$K_{p_1, \ldots, p_{k-u}}(\boldsymbol{\lambda}) = \hat{F}(\underbrace{\boldsymbol{\lambda}, \ldots, \boldsymbol{\lambda}}_{\leftarrow\, u\, \rightarrow}, \mathbf{e}(p_1), \ldots, \mathbf{e}(p_{k-u})), \tag{11.35}$$

where u takes all values from 1 to $k-1$, and $p_1, \ldots, p_{k-u}$ all the values from 1 to $g+1$. The total number of forms (11.35) is less than $k(g+1)^k$ and each is of degree $\leq k-1$ in $\boldsymbol{\lambda}$. Each form (11.35) has

$$|K_{p_1, \ldots, p_{k-u}}(\boldsymbol{\lambda})| = |F|.$$

So, by the part of Theorem 11.1 which we already know, we see that there exists $c_9(k, r, g, E)$ with the property that if $s \geq c_9$, then there is a point $\boldsymbol{\lambda}_0 \neq \mathbf{0}$ in J^s with

$$\|\boldsymbol{\lambda}_0\| \leq T$$

and

$$|K_{p_1, \ldots, p_{k-u}}(\boldsymbol{\lambda}_0)| \ll T^{-E}|F| \tag{11.36}$$

for all forms (11.35). We may suppose without loss of generality that $\boldsymbol{\lambda}_0, \mathbf{e}(1), \ldots, \mathbf{e}(g)$ are linearly independent. Writing

$$G(\mu, \nu_1, \dots, \nu_g) = F(\mu\lambda_0 + \nu_1\mathbf{e}(1) + \dots + \nu_g\mathbf{e}(g)),$$

we have $F \to G$ and

$$\Psi(F, G) \leq \max(\|\boldsymbol{\lambda}_0\|, \|\mathbf{e}(1)\|, \dots, \|\mathbf{e}(g)\|) \leq T.$$

Here G is of the type (11.34) with

$$\sigma = F(\boldsymbol{\lambda}_0), \qquad F_1(\nu_1, \dots, \nu_g) = F(\nu_1\mathbf{e}(1) + \dots + \nu_g\mathbf{e}(g))$$

and

$$H(\mu, \nu_1, \dots, \nu_g) =$$

$$\sum_{u=1}^{k-1} \binom{k}{u} \hat{F}(\underbrace{\mu\boldsymbol{\lambda}_0, \dots, \mu\boldsymbol{\lambda}_0}_{u}, \underbrace{\nu_1\mathbf{e}(1) + \dots + \nu_g\mathbf{e}(g), \dots, \nu_1\mathbf{e}(1) + \dots + \nu_g\mathbf{e}(g)}_{k-u}).$$

It follows from (11.36) that

$$|H| \ll T^{-E}|F|,$$

and the lemma is proved. □

Lemma 11.7. *Let $k \geq 1$, $t \geq 1$, $E > 0$ and $T \geq 1$. If F is a form of degree k with $s(F) \geq c_{10}(k, r, t, E)$, then*

$$F \to \sigma_1\mu_1^k + \dots + \sigma_t\mu_t^k + H_t(\mu_1, \dots, \mu_t) = G_t, \tag{11.37}$$

say, where

$$\Psi(F, G_t) \leq T \qquad \text{and} \qquad |H_t| \ll T^{-E}|F|.$$

This is the special case $g = 1$ of the following:

Lemma 11.8. *Let $k \geq 1$, $g \geq 1$, $t \geq 1$, $E > 0$ and $T \geq 1$. If F is a form of degree k with $s(F) \geq c_{11}(k, r, g, t, E)$, then there is a form G_t with $F \to G_t$ and*

$$\Psi(F, G_t) \leq T,$$

and with G_t being a form in $g + t$ variables of the type

$$G_t = \sigma_1\mu_1^k + \dots + \sigma_t\mu_t^k + F_t(\nu_1, \dots, \nu_g) + H_t(\mu_1, \dots, \mu_t, \nu_1, \dots, \nu_g),$$

where

$$|H_t| \ll T^{-E}|F|. \tag{11.38}$$

Proof. The case $t = 1$ is Lemma 11.6. From the case $t - 1$ we see that if $s \geq c_{11}(k, r, h, t - 1, 3k + 3E)$, then $F \to G_{t-1}$ with

$$G_{t-1} = \sigma_1\mu_1^k + \cdots + \sigma_{t-1}\mu_{t-1}^k + F_{t-1}(\tau_1, \ldots, \tau_h) + H_{t-1}(\mu_1, \ldots, \mu_{t-1}, \tau_1, \ldots, \tau_h), \tag{11.39}$$

$$\Psi(f, G_{t-1}) \ll T^{1/3} \tag{11.40}$$

and

$$|H_{t-1}| \ll T^{-k-E}|F|.$$

From (11.40) we get

$$|G_{t-1}| \ll \Psi(F, G_{t-1})^k|F| \ll T^{k/3}|F|,$$

so that also

$$|F_{t-1}| \ll T^{k/3}|F|,$$

where the constants implicit in $\ll$ may depend on $s(F)$.

If $h \geq c_9(k, r, g, 3k+3E)$, then, by Lemma 11.6, we have $F_{t-1} \to G$ where

$$\Psi(F_{t-1}, G) \leq T^{1/3}$$

with

$$G = \sigma_t\mu_t^k + F_t(\nu_1, \ldots, \nu_g) + H(\mu_t, \nu_1, \ldots, \nu_g) \tag{11.41}$$

and

$$|H| \ll T^{-k-E}|F_{t-1}| \ll T^{-E}|F|.$$

We may write $G(\mu_t, \nu_1, \ldots, \nu_g) = F_{t-1}(\mu_t\boldsymbol{\lambda}_0 + \nu_1\boldsymbol{\lambda}_1 + \cdots + \nu_g\boldsymbol{\lambda}_g)$ with

$$\|\boldsymbol{\lambda}_i\| = \Psi(F_{t-1}, G) \leq T^{1/3}, \qquad 0 \leq i \leq g,$$

where $\boldsymbol{\lambda}_0, \ldots, \boldsymbol{\lambda}_g$ are points in J^h. Put

$$G_t(\mu_1, \ldots, \mu_t, \nu_1, \ldots, \nu_g) = G_{t-1}(\mu_1, \ldots, \mu_{t-1}, \mu_t\boldsymbol{\lambda}_0 + \nu_1\boldsymbol{\lambda}_1 + \cdots + \nu_g\boldsymbol{\lambda}_g); \tag{11.42}$$

this makes sense because G_{t-1} has $t-1+h$ variables. We have

$$\Psi(F, G_t) \ll \Psi(F, G_{t-1})\Psi(G_{t-1}, G_t) \ll T^{2/3} < T$$

if T is large. Further, from (11.39), (11.41) and (11.42) we get

$$\begin{aligned} G_t &= \sigma_1\mu_1^k + \cdots + \sigma_{t-1}\mu_{t-1}^k + F_{t-1}(\mu_t\boldsymbol{\lambda}_0 + \nu_1\boldsymbol{\lambda}_1 + \cdots + \nu_g\boldsymbol{\lambda}_g) \\ &\qquad + H_{t-1}(\mu_1, \ldots, \mu_{t-1}, \mu_t\boldsymbol{\lambda}_0 + \nu_1\boldsymbol{\lambda}_1 + \cdots + \nu_g\boldsymbol{\lambda}_g) \\ &= \sigma_1\mu_1^k + \cdots + \sigma_{t-1}\mu_{t-1}^k + G(\mu_t, \nu_1, \ldots, \nu_g) \\ &\qquad + H_{t-1}(\mu_1, \ldots, \mu_{t-1}, \mu_t\boldsymbol{\lambda}_0 + \nu_1\boldsymbol{\lambda}_1 + \cdots + \nu_g\boldsymbol{\lambda}_g) \\ &= \sigma_1\mu_1^k + \cdots + \sigma_t\mu_t^k + F_t(\nu_1, \ldots, \nu_g) + H_t(\mu_1, \ldots, \mu_t, \nu_1, \ldots, \nu_g), \end{aligned}$$

where

$$H_t(\mu_1,\ldots,\mu_t,\nu_1,\ldots,\nu_g) = H(\mu_t,\nu_1,\ldots,\nu_g) \\ + H_t(\mu_1,\ldots,\mu_{t-1},\mu_t\boldsymbol{\lambda}_0 + \nu_1\boldsymbol{\lambda}_1 + \cdots + \nu_g\boldsymbol{\lambda}_g).$$

Here

$$|H_t| \ll |H| + \Psi(F_{t-1},G)^k|H_{t-1}| \ll T^{-E}|F| + T^{k/3-k-E}|F| \ll T^{-E}|F|.$$

To summarise: Setting $h = c_9(k,r,g,3k+3E)$ and $s = s(F) = c_{11}(k,r,h,t-1,3k+3E) = c_{11}(k,r,g,t,E)$, the assertion holds with a constant in (11.38) which depends on $k,\mathbb{K},g,t,E$ privided that T is large, as a function of these parameters. It is clear that the assertion holds for $T \geq 1$, at the possible cost of making the constant in (11.38) larger. It is also clear that the assertion is true for larger values of s, with a constant in (11.38) depending on $k,\mathbb{K},g,t$ and E, but not on s. The lemma is proved. □

Set $t = qc_3(k,r,3k+3E)$ and let F be a form with $s(F) \geq c_{10}(k,r,t,3k+3E)$. For $T \geq 1$, there is a form G_t with (11.37) and with

$$\Psi(F,G_t) \leq T^{1/3}, \qquad |H_t| \ll T^{-k-E}|F|.$$

Here

$$|G_t| \ll \Psi(F,G_t)^k|F| \ll T^{k/3}|F|,$$

so that also the form

$$A(\boldsymbol{\mu}) = \sigma_1\mu_1^k + \cdots + \sigma_t\mu_t^k$$

has

$$|A| \ll T^k|F|.$$

By Lemma 11.1 and our choice of t, there are q nonzero points

$$\boldsymbol{\mu}(i) = (\underbrace{0,\ \ldots\ ,0}_{\leftarrow(i-1)c_3\rightarrow},\boldsymbol{\nu}(i),\underbrace{0,\ \ldots\ ,0}_{\leftarrow(q-i)c_3\rightarrow}), \qquad 1 \leq i \leq q,$$

with $\boldsymbol{\nu}(i) \in J^{c_3}$, $1 \leq i \leq q$ and

$$\|\boldsymbol{\mu}(i)\| \leq T^{1/3} \qquad \text{and} \qquad |A(\boldsymbol{\mu}(i)| \ll T^{-k-E}|A| \ll T^{-E}|F|.$$

We have

$$\left|\hat{A}\big(\boldsymbol{\mu}(i_1),\ldots,\boldsymbol{\mu}(i_k)\big)\right| \ll T^{-E}|F|, \qquad 1 \leq i_1,\ldots,i_k \leq q,$$

since the left hand side is zero unless $i_1 = \cdots = i_k$. On the other hand,

$$\left|\hat{H}_t\big(\boldsymbol{\mu}(i_1),\ldots,\boldsymbol{\mu}(i_k)\big)\right| \ll T^{k/3}|H_t| \ll T^{-E}|F|, \qquad 1 \leq i_1,\ldots,i_k \leq q.$$

In other words,

$$\left|\hat{G}_t\big(\boldsymbol{\mu}(i_1),\ldots,\boldsymbol{\mu}(i_k)\big)\right| \ll T^{-E}|F|, \qquad 1 \leq i_1,\ldots,i_k \leq q.$$

Write

$$\boldsymbol{\mu}(i) = \big(\mu_1(i),\ldots\mu_t(i)\big)$$

and set

$$\boldsymbol{\lambda}(i) = \mu_1(i)\boldsymbol{\gamma}_1 + \cdots + \mu_t(i)\boldsymbol{\gamma}_t, \qquad 1 \le i \le q,$$

where

$$G_t(\boldsymbol{\mu}) = F(\mu_1\boldsymbol{\gamma}_1 + \cdots + \mu_t\boldsymbol{\gamma}_t)$$

and

$$\Psi(F, G_t) = \max \|\boldsymbol{\gamma}_i\|.$$

Then $\boldsymbol{\lambda}(i)\,(1 \le i \le q)$ are linearly independent. In fact it follows from

$$0 = \sum_{i=1}^{q} \alpha_i \boldsymbol{\lambda}(i) = \sum_{i=1}^{q} \alpha_i \sum_{j=1}^{t} \mu_j(i)\boldsymbol{\gamma}_j = \sum_{j=1}^{t} \boldsymbol{\gamma}_j \sum_{i=1}^{q} \alpha_i \mu_j(i)$$

and the linear independence of $\boldsymbol{\gamma}_j$ that

$$\sum_{i=1}^{q} \alpha_i \mu_j(i) = 0, \qquad 1 \le j \le t,$$

that is,

$$\sum_{i=1}^{q} \alpha_i \boldsymbol{\mu}(i) = 0,$$

and thus $\alpha_i = 0$, $1 \le i \le q$. The assertion follows. Therefore

$$\begin{aligned} \left|\hat{F}(\boldsymbol{\lambda}(i_1), \ldots, \boldsymbol{\lambda}(i_k))\right| &= \left|\hat{G}_t(\boldsymbol{\mu}(i_1), \ldots, \boldsymbol{\mu}(i_k))\right| \\ &\ll T^{-E}|F|, \qquad 1 \le i_1, \ldots, i_k \le q \end{aligned}$$

and

$$\|\boldsymbol{\lambda}(i)\| \ll \Psi(F, G_t)\, \|\boldsymbol{\mu}(i)\| \ll T^{2/3} < T$$

if T is large. This proves Theorem 11.1 for $h = 1$.

11.8 Proof of Theorem 11.1

We finally prove Theorem 11.1 for h forms $F_1, \ldots, F_h$ of respective degrees $k_1, \ldots, k_h$, and with $k = \max k_i$. The case $h = 1$ was established in the previous section, and we prove the theorem by induction on h. By the case $h - 1$, we can choose

$$u = c_1(k_2, \ldots, k_h, r, q, 3k + 3E).$$

Set $s = c_1(k_1, r, u, 3k + 3E)$. Then there are u linearly independent points $\boldsymbol{\nu}(1), \ldots, \boldsymbol{\nu}(u)$ of J^s with

$$\|\boldsymbol{\nu}(i)\| \le T^{1/3}, \qquad 1 \le i \le u$$

and

$$\left|\hat{F}_1(\boldsymbol{\nu}(i_1), \ldots, \boldsymbol{\nu}(i_{k_1}))\right| \ll T^{-k-E}|F|, \qquad 1 \le i_i, \ldots, i_{k_1} \le u.$$

Define new forms

$$G_j(\mu_1,\ldots,\mu_u) = F_j\big(\mu_1\boldsymbol{\nu}(1)+\cdots+\mu_u\boldsymbol{\nu}(u)\big), \qquad 1\le j\le h.$$

Then

$$|G_1| \ll T^{-k-E}|F_1|, \qquad |G_j| \ll T^k|F_j|, \quad 2\le j\le k. \tag{11.43}$$

By our choice of u, there are q linearly independent points of J^u

$$\boldsymbol{\mu}(i) = \big(\mu_1(i),\ldots,\mu_u(i)\big), \qquad 1\le i\le q$$

with

$$\|\boldsymbol{\mu}(i)\| \le T^{1/3}, \qquad 1\le i\le q$$

and

$$\big|\hat{G}_j\big(\boldsymbol{\mu}(i_1),\ldots,\boldsymbol{\mu}(i_{k_j})\big)\big| \ll T^{-k-E}|G_j| \ll T^{-E}|F_j|,$$

for $2\le j\le h$ and $1\le i_1,\ldots,i_{k_j}\le q$. On the other hand, by (11.43),

$$\big|\hat{G}_1\big(\boldsymbol{\mu}(i_1),\ldots,\boldsymbol{\mu}(i_{k_1})\big)\big| \ll T^{-E}|F_1|, \qquad 1\le i_1,\ldots,i_{k_1}\le q.$$

Put

$$\boldsymbol{\lambda}(i) = \mu_1(i)\boldsymbol{\nu}(1)+\cdots+\mu_u(i)\boldsymbol{\nu}(u), \qquad 1\le i\le q.$$

Then $\boldsymbol{\lambda}(i)$ are linearly independent

$$\|\boldsymbol{\lambda}(i)\| \ll T^{2/3} < T, \qquad 1\le i\le q$$

if T is large, and

$$\big|\hat{F}_j\big(\boldsymbol{\lambda}(i_1),\ldots,\boldsymbol{\lambda}(i_{k_j})\big)\big| = \big|\hat{G}_j\big(\boldsymbol{\mu}(i_1),\ldots,\boldsymbol{\mu}(i_{k_j})\big)\big| \ll T^{-E}|F_j|$$

for $1\le i\le h$ and $1\le i_1,\ldots,i_{k_j}\le q$. The theorem is proved. □

Notes

Theorem 11.1 (Wang [3]) is an analogue of the following theorem by Schmidt: Given positive integers h,q and odd numbers $k_1,\ldots,k_h$ and given a positive number E, however large, there is a constant $c_{12}(k_1,\ldots,k_h,q,E)$ with the following properties. Let $T\ge 1$ and $F_1(\mathbf{x}),\ldots,F_h(\mathbf{x})$ be forms with real coefficients of respective degrees $k_1,\ldots,k_h$ in $\mathbf{x}=(x_1,\ldots,x_s)$ where $s\ge c_{12}$. Then there are q linearly independent points $\mathbf{x}(1),\ldots,\mathbf{x}(q)$ in $\mathbb{Z}^s$ with $\|\mathbf{x}(i)\|\le T$ $(1\le i\le q)$ and

$$\big|\hat{F}_j\big(\mathbf{x}(i_1),\ldots,\mathbf{x}(i_{k_j})\big)\big| \ll T^{-E}|F_j|, \qquad 1\le j\le j,\ 1\le i_1,\ldots,i_{k_j}\le q.$$

The following result on diophantine equations was then derived: Given positive integers h,q and odd numbers $k_1,\ldots,k_h$ together with a positive number ϵ, however samll, there is a constant $c_{13}(k_1,\ldots,k_h,q,\epsilon)$ with the following properties. Let $G_1(\mathbf{x}), \ldots, G_h(\mathbf{x})$ be forms in $\mathbf{x} = (x_1, \ldots, x_s)$, where $S \geq c_{13}$, with respective degrees $k_1, \ldots, k_h$ and with coefficients in $\mathbb{Z}$. Then $G_1(\mathbf{x}), \ldots, G_h(\mathbf{x})$ vanish on a q-dimensional subspace spanned by q points $\mathbf{x}(1), \ldots, \mathbf{x}(q)$ in $\mathbb{Z}^s$ having $\|\mathbf{x}(i)\| \ll G^\epsilon$, $1 \leq i \leq q$. See Schmidt [3].

For any given positive integer M, let $R(M) = \{1, 2, \ldots, M\}$ denote a complete residue system modulo M. If the coefficients of $G_1(\mathbf{x}), \ldots, G_h(\mathbf{x})$ are replaced by the corresponding integers in $R(M)$, then we derive the following result: If $s \geq c_{13}$, then the congruences

$$G_i(\mathbf{x}) \equiv 0 \pmod{M}, \qquad 1 \leq i \leq h$$

has a solution $\mathbf{x} \in R(M)^s$ satifying $\|\mathbf{x}\| \ll M^\epsilon$.

Now we state the results on the problem for small solutions of congruences on forms with even degrees. Let $Q(\mathbf{x})$ be a quadratic form with coefficients in $\mathbb{Z}$. Using Minkowski's linear form theorem, Schinzel, Schlickewei and Schmidt [1] proved that the congruence

$$Q(\mathbf{x}) \equiv 0 \pmod{M} \tag{11.44}$$

has a solution $\mathbf{x} \in R(M)^s$ satisfying

$$\|\mathbf{x}\| \ll M^{\frac{1}{2}+\epsilon} \tag{11.45}$$

provided that $s \geq c_{14}(\epsilon)$, where $c_{14}(\epsilon) = O(\epsilon^{-1})$. Take $Q(\mathbf{x}) = x_1^2 + \cdots + x_s^2$. Then any nontrivial solution of (11.44) satisfies $s\|\mathbf{x}\|^2 \geq M$. This shows that the order in (11.45) is best possible. Baker [1] generalised the above result to a system of h quadratic congruences by the use of a discrete version of Schmidt's method. See Notes in Chapter 9: Let $Q_i(\mathbf{x})\,(1 \leq i \leq h)$ be h quadratic forms with coefficients in $\mathbb{Z}$. If $s \geq c_{15}(h, \epsilon)$, then

$$Q_i(\mathbf{x}) \equiv 0 \pmod{M}, \qquad 1 \leq i \leq h$$

has a solution $\mathbf{x} \in R(M)^s$ satisfying (11.45). Let $k \geq 2$ and

$$D_i(\mathbf{x}) = a_{i1}x_1^k + \cdots + a_{is}x_s^k, \qquad 1 \leq i \leq h$$

be h additive forms with coefficients in $\mathbb{Z}$. Baker also proved that the congruences

$$D_i(\mathbf{x}) \equiv 0 \pmod{M}, \qquad 1 \leq i \leq h$$

has a solution $x \in R(M)^s$ satisfying $\|\mathbf{x}\| \ll M^{\frac{1}{k}+\epsilon}$ provided that $s \geq c_{16}(k, h, \epsilon)$; see Baker [1]. However, Schmidt [4] has solved the general case of the problem for small solutions of congruences with many variables: Let $F_i(\mathbf{x})\,(1 \leq i \leq h)$ be h forms with coefficients in $\mathbb{Z}$ and with respective degrees $k_1, \ldots, k_h$. Then, if $s \geq c_{17}(k_1, \ldots, k_h, \epsilon)$, the congruences

$$F_i(\mathbf{x}) \equiv 0 \pmod{M}, \qquad 1 \leq i \leq h$$

have a solution $\mathbf{x} \in R(M)^s$ satisfying (11.45).

Using real analysis, Heath-Brown [2] proved the following theorem: Let p be a prime number and $Q(\mathbf{x})$ be a quadratic form with coefficients in $\mathbb{Z}$. Then the congruence

$$Q_i(\mathbf{x}) \equiv 0 \pmod{p}, \tag{11.46}$$

has a solution $\mathbf{x} \in R(p)^s$ satifying

$$\|\mathbf{x}\| \ll p^{\frac{1}{2}} \log p$$

provided that $s \geq 4$. We may show that the condition $s \geq 4$ is best possible because, if $s = 3$, then we may give a $Q(\mathbf{x})$ such that any solution $\mathbf{x} \in R(p)^3$ of (11.46) satisfies $\|\mathbf{x}\| \gg p^{\frac{2}{3}}$. Heath-Brown's theorem was generalised by Wang Yuan [5] to all finite fields. Recently Cochrane [2] has succeeded in improving the estimate (11.47) to

$$\|\mathbf{x}\| \ll p^{\frac{1}{2}}, \tag{11.48}$$

which is best possible apart from the improvement on the implied constant.

The problem of small solutions of congruences in algebraic number fields was first studied by Cochrane [1] who obtained the following result: Let $Q_i(\boldsymbol{\lambda}))\,(1 \leq i \leq h)$ be h quadratic forms with coefficients in J and let $\mathfrak{a}$ be an integral ideal of $\mathbb{K}$. Then the congruences

$$Q_i(\boldsymbol{\lambda}) \equiv 0 \pmod{\mathfrak{a}}, \qquad 1 \leq i \leq h$$

have a solution $\boldsymbol{\lambda} \in J^s$ satisfying

$$0 < \max_i |N(\lambda_i)| \ll N(\mathfrak{a})^{\frac{h}{h+1}+\epsilon} \tag{11.49}$$

provided that $s \geq c_{18}(h, \epsilon)$, where $c_{18} = O(\epsilon^{-1})$. Here $h/(h+1)$ may be replaced by $1/2$ when $h = 1, 2$. However, using a similar discrete version of the method given in Chapter 9, we may generalise the above two results of Baker to any algebraic number field. Thus, if $s \geq c_{19}(h, n, \epsilon)$, then the right-hand side of (11.49) can be replaced by $N(\mathfrak{a})^{\frac{1}{2}+\epsilon}$, and there is a solution $\boldsymbol{\lambda} \in J^s$ of the additive congruences

$$\sum_{j=1}^{s} \alpha_{ij} \lambda_j^k \equiv 0 \pmod{\mathfrak{a}}, \qquad 1 \leq i \leq h, \ \alpha_{ij} \in J$$

satisfying

$$0 < \max_i |N(\lambda_i)| \ll N(\mathfrak{a})^{\frac{1}{k}+\epsilon}, \quad \text{if} \quad s \geq c_{20}(k, h, n, \epsilon);$$

see Wang [6]. When $\mathbb{K}$ is a purely imaginary algebraic number field, the following result can be derived from Theorem 11.2: If $G_i(\boldsymbol{\lambda})\,(1 \leq i \leq h)$ are h forms with coefficients in J and with respective degrees $k_1, \ldots, k_h$, then, for $s \geq c_{21}(k_1, \ldots, k_h, r, \epsilon)$, the congruences

$$G_i(\boldsymbol{\lambda}) \equiv 0 \pmod{\mathfrak{a}}, \qquad 1 \le i \le h$$

have a solution $\boldsymbol{\lambda} \in J^s$ such that

$$0 < \max_i |N(\lambda_i)| \ll N(\mathfrak{a})^\epsilon.$$

Theorem 11.2 is also an improvement on a result due to Peck [1]. He first established, by a combination of the methods of Brauer [1] and Siegel [3,4], the existence of $\boldsymbol{\lambda}(i)\,(1 \le i \le q)$ with $c_{22}(k_1, \ldots, k_h, \mathbb{K}, q)$ instead of c_2, but the estimation of $\|\boldsymbol{\lambda}(i)\|$ was not considered.

This book was processed by Dr. P. Shiu using the TEX macro package from Springer-Verlag.

Reference I

R. G. Ayoub, *An Introduction to the Analytic Theory of Numbers*, Amer. Math. Soc., 1963.

R. C. Baker, *Diophantine Inequalities*, Oxford, Clarendon Press, 1986.

N. G. Chudakov, *Introduction to the Theory of Dirichlet's L-Functions*, GITTL Moscow, 1947.

H. Davenport, *Analytic Methods for Diophantine Equations and Diophantine Inequalities*, Ann Arbor Publishers, 1962.

H. Davenport, *Collected Works*, Academic Press, 1977.

T. Estermann, *Introduction to Modern Prime Number Theory*, Camb. Tracts 41, 1952.

G. H. Hardy, *Collected Papers*, Clarendon Press, Oxford, 1966.

E. Hecke, *Lectures on Theory of Algebraic Numbers*, Akad. Verlag, 1923; Springer-Verlag, 1980.

L.-K. Hua, *Introduction to Number Theory*, Science Press, Beijing, 1957; Springer-Verlag, 1982.

L.-K. Hua, *Additive Theory of Prime Numbers*, Trud. Inst. Mat. Steklov, 22, 1947; Science Press, Beijing, 1952; Amer. Math. Soc., 1965.

L.-K. Hua, *Die Abschätzung von Exponential Summen und ihre Anwendung in der Zahlentheorie*, Enz. der Math. Wiss; Heft 13, T 1, Leipzig, teubner, 1959; Science Press, Beijing, 1963.

L.-K. Hua, *Selected Papers*, Springer-Verlag, 1983.

A. A. Karatsuba, *Basic Analytic Number Theory*, Nauk, Moscow, 1975.

E. Landau, *Vorlesungen über Zahlentheorie*, Leipzig Verlag, 1927.

E. Landau, *Über einige neuere Fortschritte der additiven Zahlentheorie*, Camb. Tracts, 35, 1937.

Min Si He, *Methods in Number Theory*, (I) Science Press, Beijing, 1958; (II) ibid, 1981.

Pan Cheng Dong & Pan Cheng Biao, *Goldbach Conjecture*, Science Press, Beijing, 1981.

K. Prachar, *Primzahlverteilung*, Springer-Verlag, 1957.

C. L. Siegel, *Gesammelte Abhandlungen*, Springer-Verlag, 1966.

R. C. Vaughan, *The Hardy-Littlewood method*, Camb. Tracts, 80, 1980.

I. M. Vinogradov, *The Method of Trigonometrical Sums in the Theory of Numbers*, Trud. Inst. Math. Steklov, 23, 1947; Interscience, N.Y. 1954; Nauk, Moscow, 1971.

I. M. Vinogradov, *Basic Variants of the Method of Trigonometrical Sums*, Nauk, Moscow, 1976.

I. M. Vinogradov, *Selected Works*, Springer-Verlag, 1985.

Y. Wang, *Goldbach Conjecture*, World Science Publication Co., 1984.

References II

R. G. Ayoub [1], "On the Waring-Siegel theorem", *Can. J. Math.*, **5**, 439–450, (1953).

R. C. Baker [1], "Small solutions of quadratic and quartic congruences", *Mathematika,* **27**, 30–45, (1980).

B. J. Birch [1], "Homogeneous forms of odd degree in a large number of variables", *Mathematika,* **4**, 102–105, (1957).

B. J. Birch [2], "Waring's problem in algebraic number fields", *Proc. Camb. Phil. Soc.*, **57**, 449–459, (1961).

B. J. Birch [3], "Waring's problem for $\wp$-adic number fields", *Acta Arith.*, **9**, 169–176, (1964).

B. J. Birch [4], "Small zeros of diagonal forms of odd degree in many variables", *Proc. Lond. Math. Soc.*, **21**, 12–18, (1970).

R. Brauer [1], "A note on systems of homogeneous algebraic equations", *Bull. Amer. Math. Soc.*, **51**, 749–755, (1945).

C. Chevalley [1], "Démonstration d'une hypthèse de M. Artin", *Abh. Math. Sem. Hamburg.*, **11**, 73–75, (1935).

S. Chowla and G. Shimura [1], "On the representation of zero by a linear combination of k-th powers", *Det. Vid. Sel. Forh.*, **36**, 169–176, (1963).

T. Cochrane [1], "Small solutons of congruences over algebraic number field", *Ill. J. Math.*, **31**, 618–625, (1987).

T. Cochrane [2], "Small zeros of quadratic forms modulo p, III", (to appear).

H. Davenport [1], "Cubic forms in sixteen variables", *Pro. Roy. Soc. Lond.*, **A 272**, 285–303, (1963).

H. Davenport and H. Heilbronn [1], "On indefinite quadratic forms in five variables", *J. Lond. Math. Soc.*, **21**, 185–193, (1946).

H. Davenport and D. J. Lewis [1], "Homogeneous additive equations", *Proc. Roy. Soc. Lond.*, **A 274**, 443–460, (1963).

H. Davenport and D. Ridout [1], "Indefinite quadratic forms", *Proc. Lond. Math. Soc.*, **9**, 544–555, (1959).

Y. Eda [1], "On the Waring problem in an algebraic number field", *Sem. on modern methods in number theory, Tokyo*, (1971).

Y. Eda [2], "On Waring's problem in algebraic number fields", *Rev. Colmbiana Math.*, **9**, 29–73, (1975).

D. R. Heath-Brown [1], "Cubic forms in 10 variables", *Proc. Lond. Math. Soc.*, **47**, 225–257, (1983).

D. R. Heath-Brown [2], "Small solutions of quadratic congrunences", *Glasgow Math. J.* **27**, 87–93, (1985).

D. R. Heath-Brown [2], "Weyl's inequality, Hua's inequality and Waring's problem", *J. London Math. Soc.*, **38**, 219–230, (1988).

D. Hilbert [1], "Beweis für die Darstellbarkeit der ganzen Zahlen durch eine feste Anzahl nter Potenzen", *Göttingen Nach*, 17–36, (1909).

L. K. Hua [1], "On exponential sums over an algebraic number field", *Can. J. Math.*, **3**, 44–51, (1951).

A. Hurwitz [1], "*Mathematische Werke*", Birkhauser Verlag, (1963).

A. A. Karatsuba [1], "On the function $G(n)$ in Waring's problem", *Izv. Mat. Nauk. USSR*, **49**, 935–947, (1985).

A. Khintchine [1], "Three pearls of number theory", *Graylock Press*, (1952).

O. Körner [1], "Erweiterter des Goldbach-Vinogradovscher Satz in beliebigen algebraischen Zahlkorper", *Math. Ann.*, **143**, 344–378, (1961).

O. Körner [2], "Über die Waringsche Problem in algebraischen Zahlkörper", *Math. Ann.*, **144**, 224–238, (1961).

O. Korner [3], "Ganze algebraische Zahlen als Summen von Polynom Werten", *Math. Ann.*, **149**, 97–704.

Y. V. Linnik [1], "An elemntary solution of Waring's problem by Schnirelman's methiod", *Mat. Sb.*, **12**, 225–230, (1943).

G. A. Margulis [1], "Formes quadratiques indèfinies et flots unipote sur les espces homogénes", *C. R. Acad. Sci. Paris, Ser. I.*, **30**, 249–253, (1987).

G. A. Margulis [2], "Discrete subgroups and ergodic theory", *"Number Theory, Trace formulas and discrete groups" in honour of A. Selberg, Acad. Press*, 377–398, (1989).

T. Mitsui [1], "On the Goldbach problem in an algebraic number field I", *J. Math. Soc. Japan.*, **12**, 290–324, (1960).

T. Mitsui [2], "On the Goldbach problem in an algebraic number field II", *J. Math. Soc. Japan.*, **12**, 325–372, (1960).

L. G. Peck [1], "Diophantine equations in algebraic number fields", *Amer. J. Math.*, **71**, 387–402, (1949).

J. Pitman [1], "Bounds for solutions of diagonal equations", *Acta Arith.*, **19**, 223–247, (1971).

G. J. Rieger [1], "Über die Anzhl der Lösungen der Diophantischen Gleichung $\xi_1\zeta_1 + \xi_2\zeta_2 = \nu$ unterhalb gewisser Schranken in albraischen Zahlkörper", *J. Reine Angew. Math.*, **211**, 54–64, (1962).

A. Schinzel, H. P. Schlickewei and W. M. Schmidt [1], "Small solutions of quadratic congruences and frational parts of quadratic forms", *Acta Arith.*, **37**, 241–248, (1980).

G. J. Reiger [2], "Elementare Lösung des Waringschen problems fur algebraische Zahlkorper mit der verallgemeinerten Linnikschen methode", *Math. Ann.*, **148**, 83–88, (1962).

W. M. Schmidt [1], "Small zeros of additive forms in many variables", *Tran. Amer. Math Soc.*, **248**, 121–133, (1979).

W. M. Schmidt [2], "Small zeros of additive forms in many variables II", *Acta Math.*, **143**, 219–232, (1979).

W. M. Schmidt [3], "Diophantine inequalities for forms of odd degree", *Adv. In Math.*, **38**, 125–151, (1980).

W. M. Schmidt [4], "Small solutions of congruences in a large number of variables", *Can. Math. Bull.*, **28**, 295–305, (1985).

W. M. Schmidt [5], "Smal solutions of congruences with prime modulus", *Diophantine analysis, Proc. Number Theory Sel., Aust. Math. Soc. Conv., (1985); London Math. Soc. Lec. Notes Ser.*, **109**, 37–66, (1986).

C. L. Siegel [1], "Additive Theorie der Zahlkorper I", *Math. Ann.*, **87**, 1–35, (1922).

C. L. Siegel [2], "Additive Theorie der Zahlkorper II", *Math. Ann.*, **88**, 184–210, (1923).

C. L. Siegel [3], "Generalization of Waring's problem to algebraic number fields", *Amer. J. Math.*, **66**, 122–136, (1944).

C. L. Siegel [4], "Sums of m-th powers of algebraic integers", *Ann. Math.*, **46**, 313–339, (1945).

R. M. Stemmler [1], "The easier Waring problem in algebraic number fields", *Acta Arith.*, **6**, 447-468, (1961).

M. V. Subbarao and Wang Yuan [1], "On a generalized Waring's problem in algebraic number fields", *in "Number Theory" in honour of L. K. Hua. Springer-Verlag*, (1990).

T. Tatuzawa [1], "On the Waring problem in an algebraic number field", *J. Math. Soc. Japan.*, **10**, 322–341, (1958).

T. Tatuzawa [2], "On the Waring problem in an algebraic number field", *Sem. On modern methods in number theory*, Tokyo (1971).

T. Tatuzawa [3], "On Waring's problem in algebraic number fields", *Acta Arith.*, **24**, 37–60, (1973).

R. C. Vaughan [1], "On Waring's problem for cubes", *J. Reine Angew.*, **365**, 122–170, (1986).

R. C. Vaughan [2], "On Waring's problem for smaller exponents II", *Mathematika*, 6–22, (1986).

R. C. Vaughan [3], "Recent work on Waring's problem", *"Number Theory, Trace formulas and discrete groups" in honour of A. Selberg, Acad. Press*, 503–510, (1989).

Y. Wang [1], "Bounds for solutions of additive equations in an algebraic number field I", *Acta Arith.*, **48**, 21–48, (1987).

Y. Wang [2], "Bounds for solutions of additive equations in an algebraic number field II", *Acta Arith.*, **48**, 307–323, (1987).

Y. Wang [3], "Diophantine inequalities for forms in an algebraic number field", *J. Number theory*, **29**, 324–344, (1989).

Y. Wang [4], "On homogeneous additive congruences", *Sci. Sinica*, **32**, 524–536, (1989).

Y. Wang [5], "On small zeros of quadratic forms over finite fields", *J. Number Theory*, **31**, 272–284, (1989).

Y. Wang [6], "Small solutions of congruences", to appear.

H. Weyl [1], "Uber die Gleichverteilung von Zahlen mod Eins", *Math. Ann.*, **77**, 313–352 (1916).

T. D. Wooley [1], "Large improvements in Waring's problem", to appear.

Index